Environmental Science Step by Step

A Structured Introduction to Environmental Systems & Earth's Ecosystems

Robert Peterson

PREFACE

Welcome to "Environmental Science Step by Step"! Whether you're a student, educator, or simply someone curious about our planet, you've picked up a book that aims to be your friendly guide through the fascinating field of environmental science.

Before I started writing this book, I asked myself: "How can we make environmental science both accessible and comprehensive?" The answer lay in breaking down complex concepts into manageable, logical steps while never losing sight of the bigger picture – the web of relationships that makes our planet function as a living, breathing system.

Why This Book?

Environmental science isn't just another subject – it's the study of our home and our future. As we face unprecedented environmental challenges, from climate change to biodiversity loss, understanding how our planet works has never been more critical. This book is designed to equip you with the knowledge, concepts, and frameworks to comprehend these challenges and, hopefully, inspire you to be part of the solution.

What Makes This Book Different?

1. **Structured Approach**: Each chapter builds upon the previous one, creating a logical progression from fundamental concepts to complex environmental issues. We start with the basics of environmental science and gradually explore more specific topics, ensuring you have a solid foundation before tackling more advanced subjects.
2. **Real-World Connections**: Throughout the book, you'll find examples and case studies that connect theoretical concepts to real-world situations. Environmental science isn't just about abstract ideas – it's about understanding the world around us and how we interact with it.
3. **Holistic Perspective**: While we break down environmental science into digestible chapters, we consistently emphasize the interconnectedness of Earth's systems. You'll learn how air quality affects water systems, how soil health impacts biodiversity, and how human activities influence all of these elements.

How to Use This Book

Think of this book as a guided journey. Start with Chapter 1, which lays the groundwork for understanding environmental science. Each subsequent chapter builds upon this foundation, introducing new concepts while reinforcing what you've already learned.

Don't rush – take time to absorb each concept. The "Step by Step" in our title isn't just for show. Environmental science can seem overwhelming at first, but by breaking it down into manageable pieces, you'll find that even complex topics become understandable.

At the same time, feel free to skip around to the parts that are most interesting to you if that fits your reading style. We've designed the book so that each chapter and individual section can stand on its own. We don't neglect to mention a concept simply because it's been mentioned earlier.

A Note on Perspective

Environmental science is a field where facts intersect with values. While this book strives to present accurate, scientifically-backed information, we also acknowledge that many environmental issues involve complex trade-offs and differing viewpoints. We encourage you to think critically, consider multiple perspectives, and form your own informed opinions.

Who Is This Book For?

This book is designed for:

- High school and college students studying environmental science
- Educators looking for a comprehensive, well-structured teaching resource
- Environmental professionals seeking a refresher or reference guide
- Anyone interested in learning about our planet and its environmental systems

Looking Ahead

As you progress through this book, you'll develop a deeper understanding of how our planet works and our role within it. You'll learn about the challenges we face, but also about the innovative solutions being developed to address them. Most importantly, you'll gain the knowledge to make informed decisions about environmental issues and perhaps be inspired to contribute to a more sustainable future.

Environmental science isn't just about problems – it's about possibilities. As you read this book, keep in mind that every bit of knowledge you gain is a step toward better understanding and protecting our remarkable planet.

So, let's begin our study together, step by step, through the discipline of environmental science. Whether this is your first encounter with the subject or you're building on existing knowledge, I hope this book serves as both an informative guide and an inspiration for further exploration.

TOPICAL OUTLINE

Chapter 1: Introduction to Environmental Science

- What is Environmental Science?
- The Interconnectedness of Earth's Systems
- Human-Environment Interactions
- The Scientific Method in Environmental Studies
- Sustainability and Sustainable Development
- Biodiversity and Its Importance
- Natural Resources: Renewable vs Non-Renewable
- Ecological Footprints
- Environmental Ethics
- Global Environmental Challenges

Chapter 2: Ecosystems and Energy Flow

- Components of Ecosystems: Biotic and Abiotic Factors
- Trophic Levels and Food Chains/Webs
- Energy Flow: The 10% Rule and Energy Loss
- Keystone Species and Ecosystem Balance
- Ecosystem Resilience and Recovery After Disturbance

Chapter 3: Biogeochemical Cycles

- The Carbon Cycle
- The Nitrogen Cycle
- The Water Cycle (Hydrological Cycle)
- Human Impact on Biogeochemical Cycles
- Sulfur Cycle and Its Environmental Effects

Chapter 4: The Atmosphere and Climate Systems

- Layers of the Atmosphere
- Earth's Climate Zones
- Atmospheric Circulation and Weather Patterns
- The Role of Oceans in Climate Regulation
- Climate Feedback Mechanisms: Positive and Negative

Chapter 5: Biodiversity and Conservation

- Levels of Biodiversity: Genetic, Species, Ecosystem
- Threats to Biodiversity: Habitat Loss, Overexploitation
- Conservation Strategies: In-Situ and Ex-Situ
- Ecosystem Services and Their Importance to Humans
- Community-Based Conservation Approaches

Chapter 13: Environmental Health and Toxicology

- Toxins in the Environment: Types and Sources
- Bioaccumulation and Biomagnification
- Human Health and Environmental Hazards
- Endocrine Disruptors and Long-Term Health Impacts

Chapter 14: Sustainable Energy

- Fossil Fuels: Impacts and Alternatives
- Renewable Energy: Solar, Wind, Hydro, Geothermal
- Energy Efficiency and Conservation
- Decentralized Energy Systems and Microgrids

Chapter 15: Waste Management

- Types of Waste: Municipal, Industrial, Hazardous
- Waste Disposal Methods: Landfills, Incineration, Recycling
- The Circular Economy and Waste Reduction
- Composting and Organic Waste Solutions
- Plastic Waste and Ocean Pollution

Chapter 16: Environmental Policy and Regulation

- Environmental Laws: National and International
- Role of Governments in Environmental Protection
- Non-Governmental Organizations and Activism
- Public Participation in Environmental Decision-Making

Appendix

- Terms and Definitions

Afterword

TABLE OF CONTENTS

CHAPTER 1: INTRODUCTION TO ENVIRONMENTAL SCIENCE

What is Environmental Science?

Environmental science is the study of the natural world and how humans interact with it. It looks at how ecosystems function and how different elements of the environment—like air, water, soil, and living organisms—work together. Environmental science is about understanding the balance between these elements and how human activities influence this balance.

The field pulls from a wide range of disciplines, including biology, chemistry, physics, and earth sciences, making it a broad, interdisciplinary study. Environmental scientists study ecosystems and biodiversity, which means they look at how living organisms, from plants to animals, interact within their environments. This includes understanding food chains, natural habitats, and the roles different species play in maintaining ecosystems. For instance, bees pollinate plants, which is vital for agriculture, while predators like wolves control prey populations, keeping ecosystems healthy.

One of the key focuses in environmental science is **examining human impact on the environment**. Industrial activities, urbanization, and deforestation are all examples of how human activities can disrupt natural systems. When forests are cleared, it leads to habitat loss for many species, and when fossil fuels are burned, it increases greenhouse gas emissions, contributing to climate change.

Another area studied is **pollution and waste management**. Pollutants, whether from factories, vehicles, or agriculture, end up in the air, water, or soil, affecting not only the environment but human health as well. Chemicals like pesticides can leach into groundwater, contaminating drinking water sources, while plastic waste accumulates in oceans, endangering marine life. Environmental science investigates the origins, impacts, and potential solutions to these problems. Scientists track pollutant levels and study how different pollutants spread through ecosystems, how long they persist, and what effects they have on both living organisms and physical environments.

Environmental science also deals with **natural resources**—renewable resources like forests, water, and wind, and nonrenewable resources such as fossil fuels and minerals. Resource management is critical to avoid overexploitation. For instance, unsustainable logging practices can lead to deforestation, while excessive water use can cause shortages, especially in arid regions. Scientists look for ways to use resources more sustainably, such as promoting renewable energy over fossil fuels to reduce environmental harm and ensure resources are available for future generations.

Another major topic is **climate change**, which affects weather patterns, sea levels, and ecosystems across the globe. Environmental science uses data from meteorology, oceanography, and geology to model and predict changes in the earth's climate system. This helps societies prepare for impacts like rising sea levels, which can displace coastal communities, or more frequent extreme weather events like hurricanes and droughts. Understanding how greenhouse gases, such as carbon dioxide and methane, contribute to the warming of the earth's atmosphere is essential for developing policies to mitigate these changes.

Biodiversity loss is another concern. As habitats are destroyed or altered, many species are at risk of extinction, which reduces biodiversity. This loss can destabilize ecosystems, as each species contributes to the overall functioning of its environment. Scientists work to identify endangered species, understand the causes of their decline, and develop strategies to protect them.

The **hydrological cycle** is another key subject. Water moves through the environment in a continuous cycle—evaporation from oceans, condensation into clouds, and precipitation as rain or snow. Environmental scientists study how this cycle works and how human activities, like dam construction or irrigation, alter it. Water quality and availability are often areas of concern, especially as the demand for freshwater increases due to population growth and agricultural needs.

Invasive species are another issue. When non-native species are introduced to an ecosystem, either accidentally or intentionally, they can outcompete local species, leading to imbalances. For instance, species like zebra mussels in North America have severely impacted water ecosystems by clogging pipes and outcompeting native species.

In environmental science, all these factors—ecosystems, human impact, natural resources, climate change, and biodiversity—are interconnected. **Every action within an ecosystem has a ripple effect**, influencing both the local environment and, often, global systems. By studying these connections, environmental scientists aim to find ways to preserve the environment while balancing human needs.

The Interconnectedness of Earth's Systems

Earth's systems are deeply interconnected, creating a complex web of interactions that shape the planet's environment. These systems—atmosphere, hydrosphere, geosphere, and biosphere—work together in dynamic ways, each influencing the others through processes such as energy transfer, chemical cycles, and biological activity.

The **atmosphere** is the layer of gases surrounding Earth, vital for sustaining life. It interacts closely with the **hydrosphere**, which includes all water on the planet, from oceans and rivers to groundwater and glaciers. For example, the water cycle is a fundamental process linking these systems. Water evaporates from the surface, condenses in the atmosphere, and returns as precipitation, regulating temperature and distributing fresh water across the globe. Changes in one part of this cycle—like increased evaporation due to warming oceans—can lead to shifts in weather patterns, showing how sensitive the system is to disturbances.

The **geosphere**, which includes the Earth's crust, mantle, and core, interacts with both the atmosphere and hydrosphere. For instance, volcanic activity releases gases like carbon dioxide and water vapor into the atmosphere, influencing the climate. Over millions of years, the slow movement of tectonic plates also shapes landscapes, forming mountains and triggering earthquakes. These geological processes affect ecosystems in the biosphere, as mountain ranges can alter wind and precipitation patterns, creating unique habitats for different species.

The **biosphere**, which encompasses all living organisms, is tightly linked with the other systems. Plants, for example, have a critical role in the carbon cycle, absorbing carbon dioxide from the atmosphere and releasing oxygen through photosynthesis. This process regulates atmospheric composition and helps maintain the balance needed for life. On the other hand, animals influence the soil and water systems. For instance, beavers create dams that change water flow and create new ecosystems, while burrowing animals aerate the soil, impacting its nutrient content.

A key aspect of this interconnectedness is the **carbon cycle**, a process that moves carbon through the atmosphere, hydrosphere, geosphere, and biosphere. Carbon is stored in rocks, oceans, and living organisms, and it cycles through these systems in various forms. Human activities, like burning fossil fuels, disrupt this natural cycle by releasing large amounts of carbon dioxide into the atmosphere, leading to global warming. This, in turn, affects the hydrosphere by warming oceans, which can lead to coral bleaching and disrupt marine ecosystems. The geosphere is impacted through processes like ocean acidification, which dissolves carbonate materials, affecting the organisms that rely on them.

The **nitrogen cycle** is another example of interconnection. Nitrogen, an essential element for life, moves between the atmosphere, soil, and living organisms. Bacteria in the soil fix nitrogen from the air, making it available for plants. In turn, animals obtain nitrogen by consuming plants, and when they die, decomposers return nitrogen to the soil. However, human activities like the overuse of fertilizers can overload this system, leading to excess nitrogen runoff into waterways, causing eutrophication and damaging aquatic ecosystems.

The **energy balance** of Earth is another key point of interaction. Solar energy drives most processes on Earth, from weather patterns to photosynthesis in plants. However, the energy received from the Sun is unevenly distributed across the planet, leading to variations in temperature, which in turn drive the movement of air and water. For example, warm air near the equator rises and moves toward the poles, while cooler air sinks, creating global wind patterns. These winds drive ocean currents, which distribute heat around the planet, linking the atmosphere and hydrosphere in a global climate system.

Disruptions in one system can cascade through others. For instance, deforestation (a change in the biosphere) affects the atmosphere by reducing the planet's ability to absorb carbon dioxide, leading to increased greenhouse gas levels and climate change. These changes alter precipitation patterns (hydrosphere), potentially leading to more extreme weather events.

Understanding the interconnectedness of Earth's systems highlights how delicate and balanced these interactions are. Changes in one system can have far-reaching effects across others, emphasizing the importance of studying these relationships to maintain a stable environment for life on Earth.

Human-Environment Interactions

Human-environment interactions are an ongoing and ever-evolving relationship, where human activities affect natural systems, and in turn, natural processes influence human societies. This interaction manifests across a wide range of activities, from agriculture and urbanization to energy production and resource extraction, shaping both the environment and human development.

Agriculture, one of the earliest forms of human-environment interaction, significantly alters ecosystems. By clearing forests for farmland, humans change the landscape, disrupting local ecosystems and biodiversity. In tropical regions, deforestation for agriculture reduces carbon sequestration, contributing to climate change. Additionally, the use of fertilizers and pesticides in farming introduces chemicals into the soil and water systems, which can lead to pollution and the disruption of local aquatic life. These chemicals can leach into rivers, causing eutrophication, where excess nutrients promote algae growth, depleting oxygen and harming fish populations.

Urbanization is another major way humans interact with their environment. As cities expand, natural land is converted into buildings, roads, and infrastructure, leading to habitat loss for many species. Urban areas also contribute to the **urban heat island effect**, where cities are significantly warmer than surrounding rural areas due to human activity and materials like asphalt absorbing heat. This temperature increase affects local weather patterns, leading to more intense storms or altered rainfall distribution. Water management becomes critical in urban areas as impermeable surfaces prevent natural absorption of rainfall, often causing flooding.

Energy production has a profound impact on the environment, with fossil fuel use being one of the most significant. Extracting and burning coal, oil, and natural gas releases large amounts of carbon dioxide and other pollutants into the atmosphere, contributing to global warming. This process also disrupts local ecosystems, especially in cases like oil spills, which can devastate marine life and coastal ecosystems. Renewable energy sources, like wind and solar, are seen as a way to reduce this impact, though they also come with their own environmental considerations. For example, large-scale wind farms can alter bird migration patterns, while solar farms require large land areas that might displace local wildlife.

Water use is another critical area of human-environment interaction. Freshwater resources are limited, and human activities such as agriculture, industry, and household consumption place increasing pressure on water availability. In some regions, over-extraction of groundwater has led to the depletion of aquifers, reducing water availability for future generations. Dams built for hydroelectric power and irrigation also alter natural water flow, affecting aquatic ecosystems and fish migration patterns.

Climate change represents a significant interaction between humans and the environment. The release of greenhouse gases from industrial activities, transportation, and agriculture has led to a warming climate, which in turn affects weather patterns, sea levels, and ecosystems. Rising sea levels, for instance, threaten coastal cities, leading to human displacement and the need for adaptive strategies like sea walls or relocation. Meanwhile, changing weather patterns can affect food security, as shifting rainfall patterns and increasing temperatures disrupt traditional farming practices.

Finally, **resource extraction**, such as mining for minerals or drilling for oil, often leads to significant environmental degradation. These activities can lead to habitat destruction, water contamination, and air pollution. Open-pit mining, for example, can completely remove ecosystems and alter landscapes. In addition, the waste products from resource extraction, such as tailings from mining operations, can contain toxic materials that leach into the soil and water, posing long-term environmental hazards.

In all these cases, the environment also impacts humans. Natural disasters like hurricanes, floods, or droughts are intensified by human-induced changes, such as climate change and deforestation. As human populations grow and consumption increases, the balance of human-environment interactions becomes more delicate, making sustainable management practices essential for the future.

The Scientific Method in Environmental Studies

The scientific method is a critical framework in environmental studies, enabling researchers to systematically investigate and understand the complex interactions within natural systems. It allows scientists to observe, hypothesize, experiment, and draw conclusions that help address environmental issues. The method ensures that findings are based on evidence and logic, reducing biases and errors in environmental research.

The first step in the scientific method is **observation**. Environmental scientists begin by observing phenomena in nature, such as changes in ecosystems, climate patterns, or pollution levels. These observations help identify potential environmental problems or patterns that need further investigation. For example, rising global temperatures observed over several decades led to questions about climate change and its causes. Similarly, the discovery of a declining population in a species could prompt an investigation into habitat loss or pollution.

Following observation, scientists develop a **research question** or identify a problem to study. This question should be clear and focused. In environmental studies, these questions often revolve around understanding how human activities impact ecosystems or identifying solutions to mitigate damage. For instance, researchers may ask, "How does urban runoff affect freshwater ecosystems?" or "What are the effects of deforestation on local biodiversity?"

Next, scientists form a **hypothesis**, a testable statement that provides a possible explanation for the observed phenomenon. In environmental studies, hypotheses often address cause-and-effect relationships. A scientist might hypothesize, for example, that increasing urbanization leads to higher levels of water pollution due to the runoff of chemicals and waste into nearby rivers.

Once a hypothesis is established, **experiments or data collection** begin. This step involves gathering evidence to test the hypothesis. In environmental studies, researchers often rely on both fieldwork and laboratory experiments. Fieldwork might involve collecting soil, water, or air samples, while laboratory work could involve testing for pollutants or studying species behavior under controlled conditions. In some cases, scientists use computer models to simulate environmental changes and predict future outcomes, such as how rising sea levels will affect coastal communities.

When conducting experiments, scientists need to control variables to ensure that the results are reliable. They also perform multiple trials to confirm their findings. For instance, if a study aims to determine how a specific pollutant affects plant growth, scientists would need to control for other factors like sunlight, water, and soil quality, which could also influence the results.

Once the data is collected, scientists **analyze** it to determine whether it supports or contradicts the hypothesis. In environmental studies, this often involves statistical analysis to understand patterns or correlations. For example, if researchers are studying the relationship between deforestation and species extinction, they may analyze historical data on forest cover and species populations to determine if there is a significant correlation.

6

If the hypothesis is supported, scientists may refine it further or explore new questions that arise from the findings. If the hypothesis is contradicted, they may revise it and conduct additional experiments to better understand the problem. This iterative process ensures that scientific conclusions are robust and well-supported by evidence.

Finally, scientists **communicate** their findings through reports, papers, or presentations. This step is crucial in environmental studies because the results often inform public policy and environmental management decisions. For example, research showing that certain chemicals are harmful to marine life might lead to regulations limiting their use. The scientific method, when applied rigorously, helps ensure that such decisions are grounded in reliable data.

Replication of studies is also important in environmental science. Other scientists must be able to reproduce the results to confirm their validity. Peer review further ensures that the research is accurate, as other experts in the field evaluate the methodology and conclusions.

The scientific method provides a structured, objective approach to studying environmental problems. It ensures that environmental science remains a reliable field where data, not speculation, drives solutions to complex issues like climate change, pollution, and biodiversity loss.

Sustainability and Sustainable Development

Sustainability and sustainable development are key concepts in environmental science, aiming to balance human needs with the planet's ability to regenerate its resources. Sustainability is the practice of using resources in a way that preserves their availability for future generations, while sustainable development seeks to promote economic and social progress without compromising the health of ecosystems.

At its core, **sustainability** focuses on three interconnected pillars: environmental protection, social equity, and economic viability. These pillars are often referred to as the "triple bottom line." For a system to be truly sustainable, it must meet the needs of the present without depleting resources for future generations. This means using natural resources at a rate at which they can be replenished, minimizing waste, and reducing environmental damage.

Environmental sustainability emphasizes conserving natural resources and protecting ecosystems. This involves practices such as reforestation, water conservation, and the use of renewable energy sources. For example, shifting from fossil fuels to solar or wind power reduces greenhouse gas emissions and limits the depletion of nonrenewable resources like coal and oil. Sustainable farming practices, such as crop rotation and organic farming, help maintain soil health and reduce the need for chemical inputs, preserving biodiversity and reducing pollution.

Social sustainability focuses on ensuring that the benefits of development are shared equitably across societies. This means addressing issues like poverty, access to clean water, healthcare, and education. Sustainable development aims to create resilient communities that can adapt to environmental challenges while maintaining a decent standard of living. For instance, providing access to clean energy in developing countries can improve quality of life while reducing reliance on polluting fuels like wood or coal. Equitable distribution of resources is central to this, ensuring that all populations, especially vulnerable groups, can participate in and benefit from sustainable practices.

Economic sustainability ensures that economic activities are viable in the long term, without causing environmental degradation. This involves creating business models that prioritize resource efficiency, waste reduction, and social responsibility. For example, companies adopting circular economy principles aim to minimize waste by recycling materials and designing products that last longer. In agriculture, sustainable practices can enhance long-term productivity by maintaining soil health, which prevents the need for costly interventions like chemical fertilizers.

One important concept within sustainability is **the precautionary principle**, which suggests that when an activity poses a potential risk to the environment or human health, precautionary measures should be taken, even if the full extent of the risk is not yet known. This principle is often applied in environmental policy, where the long-term impacts of certain practices, like the widespread use of plastics or pesticides, are not fully understood. By exercising caution, societies can avoid irreversible damage to ecosystems.

Sustainable development also addresses the **finite nature of many resources**. For example, fossil fuels like oil and natural gas are nonrenewable, meaning they cannot be replenished on a human timescale. Once depleted, they are gone. Therefore, sustainable development promotes the use of renewable resources, such as wind, solar, and hydro energy, which can be naturally regenerated.

Sustainable urban planning is another critical aspect of sustainable development. As the global population grows and more people move to cities, urban areas must be designed to reduce environmental impact. This includes creating green spaces, reducing energy consumption, promoting public transportation, and developing infrastructure that supports sustainable living. Green buildings, which use energy efficiently and minimize waste, are examples of how urban design can contribute to sustainability.

Sustainability and sustainable development require global cooperation, as environmental issues like climate change, deforestation, and resource depletion affect all nations. International agreements, such as the Paris Agreement on climate change, represent efforts to promote sustainability on a global scale.

Biodiversity and Its Importance

Biodiversity refers to the variety of life on Earth, encompassing all species of plants, animals, fungi, and microorganisms, as well as the genetic differences within these species and the ecosystems they form. This diversity is critical to the health and functioning of ecosystems, as it supports the complex interactions that allow life to thrive.

One of the key reasons biodiversity is important is because it ensures **ecosystem stability and resilience**. In ecosystems, different species perform various roles—such as pollinators, decomposers, or predators—that keep the system balanced. For example, bees pollinate plants, which are essential for food production, while fungi decompose dead organic matter, returning nutrients to the soil. If one species is lost, others can often take over its role, but a system with lower biodiversity is more vulnerable to disruption. When biodiversity is high, ecosystems are more adaptable to changes, such as shifts in climate or the introduction of invasive species.

Biodiversity also is important in **ecosystem services**, which are the natural processes that benefit humans. These include things like clean air, water filtration, soil fertility, and pollination. Forests, for instance, act as carbon sinks, absorbing carbon dioxide from the atmosphere and helping mitigate climate change. Wetlands filter pollutants from water, maintaining water quality, while coastal mangroves protect against storm surges and erosion. A decline in biodiversity can weaken these services, threatening the resources humans depend on for survival.

Another critical aspect of biodiversity is its **contribution to food security**. The variety of species in agriculture, known as agrobiodiversity, helps ensure that crops and livestock can survive in different conditions. For example, certain crop varieties are more resistant to pests or drought, while genetic diversity in livestock can reduce the risk of disease wiping out entire populations. Maintaining biodiversity in agriculture allows humans to adapt to environmental challenges, such as changing climate patterns or new agricultural pests.

Medicinal resources are another area where biodiversity proves vital. Many modern medicines are derived from compounds found in plants, fungi, and microorganisms. For example, the cancer treatment drug paclitaxel was first extracted from the bark of the Pacific yew tree. The loss of biodiversity could mean the extinction of species that hold the key to curing diseases. Conservation of diverse ecosystems ensures that future generations can benefit from the natural pharmacy provided by nature.

Biodiversity is also tied to **cultural and economic values**. Many indigenous communities rely on biodiversity for their livelihoods, food, and traditional medicine. Moreover, ecotourism, which focuses on experiencing natural environments, is a growing industry that depends on preserving wildlife and ecosystems. The economic benefits of biodiversity extend beyond just ecotourism, though. Healthy ecosystems support industries like agriculture, forestry, and fisheries, which are essential to the global economy.

Human activities, such as deforestation, pollution, overfishing, and climate change, are threatening biodiversity at an unprecedented rate. **Habitat loss** is one of the most significant drivers of species extinction. As forests are cleared for agriculture or urban development, species lose their homes and food sources, leading to population declines. Climate change exacerbates this by altering habitats and forcing species to migrate to survive.

The protection of biodiversity is essential not only for preserving nature but for maintaining the balance that supports life on Earth. Through conservation efforts, such as protected areas, wildlife corridors, and sustainable development, humans can work to safeguard the biodiversity that underpins ecosystem stability and human well-being.

Natural Resources: Renewable vs Non-Renewable

Natural resources are materials and substances found in nature that humans use to meet their needs and support economic development. These resources are generally categorized into two types: renewable and non-renewable. Understanding the difference between these two categories and how they are used is essential for developing sustainable practices and managing the Earth's resources wisely.

Renewable resources are those that can naturally replenish over time. They are either continuously available, like sunlight and wind, or can regenerate relatively quickly, such as forests and freshwater. These resources are often considered sustainable because, with proper management, they can be used without being depleted.

One of the most abundant renewable resources is **solar energy**. Solar power is generated by converting sunlight into electricity using photovoltaic cells. It is a clean and virtually limitless resource that does not produce greenhouse gases, making it an essential tool in the fight against climate change. Similarly, **wind energy** is harnessed through wind turbines that convert the kinetic energy of wind into electricity. Wind farms, which are often located in open plains or offshore, provide a renewable and low-impact source of power.

Hydropower, generated by capturing the energy of moving water, is another renewable resource. Dams and other structures are used to control water flow and generate electricity. While hydropower is considered renewable, it does come with environmental challenges. Large dams can disrupt aquatic ecosystems, affect fish migration, and alter river flows. Managing these impacts is critical to ensuring hydropower remains a sustainable resource.

Forests are another renewable resource that provides materials like timber and fuel while supporting biodiversity and acting as carbon sinks. However, forests must be managed sustainably to prevent deforestation. When trees are harvested faster than they can regrow, forest ecosystems collapse, leading to habitat loss and increased carbon emissions. Sustainable forestry practices, such as selective logging and replanting, help maintain forest resources over the long term.

Non-renewable resources, on the other hand, are finite. These are resources that take millions of years to form and cannot be replaced once they are extracted and used. The most prominent examples of non-renewable resources are fossil fuels, including coal, oil, and natural gas.

Fossil fuels are formed from the remains of ancient plants and animals buried under layers of sediment for millions of years. When burned, they release energy that powers much of the modern world. However, the extraction and use of fossil fuels come with significant environmental costs. Burning fossil fuels produces greenhouse gases like carbon dioxide, which contribute to global warming. Oil spills, coal mining, and natural gas fracking also lead to habitat destruction, water contamination, and air pollution.

Minerals are another type of non-renewable resource. These include metals like copper, aluminum, and iron, as well as rare earth elements used in technology. Mining these materials can have devastating environmental effects, including deforestation, soil degradation, and water pollution. As demand for these materials grows—particularly for use in electronics and renewable energy technologies—the challenge becomes finding ways to extract and use them more sustainably.

The key difference between renewable and non-renewable resources is the rate at which they can be replenished. Renewable resources, such as wind or solar energy, are virtually inexhaustible if managed well, while non-renewable resources like fossil fuels and minerals are finite and will eventually run out. The **sustainability** of resource use depends not only on the type of resource but also on how it is managed.

For example, overuse of **renewable resources** like freshwater or forests can lead to depletion, just as unsustainable as the extraction of non-renewable resources. Water shortages in some parts of the world are due not to a lack of available freshwater but to over-extraction and poor management. Similarly, renewable resources like soil can degrade if not properly cared for, through practices such as over-farming or deforestation.

Transitioning to **renewable energy sources** is a key part of moving toward sustainability, but this requires careful planning to manage both renewable and non-renewable resources efficiently. While non-renewable resources are still critical for many industries, reducing their use and developing recycling technologies can help extend their availability and lessen their environmental impact.

Ecological Footprints

An ecological footprint measures the impact of human activities on the environment in terms of the amount of natural resources consumed and the waste produced. It quantifies how much land and water area is required to sustain the consumption and waste disposal of an individual, community, or nation. The larger the ecological footprint, the greater the environmental impact.

The concept was developed to provide a clear way to understand the environmental pressures created by human activities. It measures six main components: carbon footprint, cropland, grazing land, fishing grounds, built-up land, and forest areas required to absorb carbon dioxide emissions. Each component assesses how much of Earth's resources are being used and whether this use exceeds the planet's ability to regenerate those resources.

A significant part of most ecological footprints is the **carbon footprint**, which represents the amount of carbon dioxide emissions generated by burning fossil fuels for energy. Energy use, especially from non-renewable sources like coal, oil, and natural gas, contributes heavily to climate change through greenhouse gas emissions. Carbon footprints can be reduced by shifting to renewable energy sources, such as solar and wind power, and improving energy efficiency.

Agricultural activities, including cropland and grazing land, are another large contributor to ecological footprints. Modern industrial agriculture often requires large amounts of water, land, and chemical inputs like pesticides and fertilizers, which can deplete soil nutrients, pollute waterways, and degrade ecosystems. The ecological footprint accounts for how much land is needed to grow food and raise livestock, as well as the impact of these practices on biodiversity and soil health.

Fishing grounds are an essential part of the ecological footprint, especially for communities that rely heavily on seafood. Overfishing and destructive fishing practices can deplete fish stocks and damage marine ecosystems, leading to long-term declines in fish populations and biodiversity loss. Sustainable fishing practices, such as catch limits and marine protected areas, are critical for keeping fish populations at healthy levels and reducing the ecological footprint of fishing.

Built-up land, such as cities, roads, and infrastructure, contributes to an ecological footprint by replacing natural habitats with urban and industrial spaces. Urbanization often leads to habitat loss, increased pollution, and higher energy consumption. Expanding cities and industries often require clearing forests, draining wetlands, and altering ecosystems, which disrupts biodiversity. Sustainable urban planning, with a focus on reducing sprawl and increasing green spaces, can help mitigate the impact of built-up land on ecosystems.

Forests are vital in absorbing carbon dioxide and maintaining biodiversity. Deforestation, primarily driven by agriculture and logging, contributes to biodiversity loss, climate change, and the depletion of natural resources. The ecological footprint calculates how much forest area is needed to offset carbon emissions and provide resources like timber. Reforestation and sustainable forest management can reduce the ecological footprint by restoring forest cover and protecting ecosystem services.

The idea behind the ecological footprint is to compare humanity's demand for resources to the Earth's capacity to regenerate those resources, often referred to as the **biocapacity**. If the footprint exceeds biocapacity, it indicates that we are using resources faster than they can be replenished, leading to environmental degradation. Currently, global ecological footprints exceed Earth's biocapacity, meaning we are in a state of **ecological overshoot**. This overuse of resources leads to problems such as deforestation, water scarcity, climate change, and loss of biodiversity.

Reducing an ecological footprint involves making more sustainable choices. On an individual level, this could mean using energy-efficient appliances, reducing meat consumption, recycling, and opting for sustainable transportation methods like walking, cycling, or using public transport. At the national and global levels, it involves transitioning to renewable energy, implementing sustainable land-use policies, and promoting conservation efforts to protect natural habitats.

Ecological footprints provide a valuable tool for understanding and addressing the environmental impacts of human consumption. By measuring how much we use and comparing it to how much the planet can provide, we can identify unsustainable practices and develop strategies to live within Earth's limits.

Environmental Ethics

Environmental ethics is a branch of philosophy that considers the moral relationships between humans and the natural environment. It asks fundamental questions about how we should treat nature and what responsibilities we have to protect ecosystems, species, and the Earth itself. The field has gained prominence as environmental problems like climate change, deforestation, and biodiversity loss have become more urgent.

At the heart of environmental ethics is the question of **intrinsic vs. instrumental value**. Traditionally, nature has been viewed primarily as having instrumental value, meaning its worth is based on the benefits it provides to humans—such as resources, recreation, and aesthetic enjoyment. In contrast, many environmental ethicists argue that nature has intrinsic value, meaning it has worth in its own right, independent of its usefulness to humans. This perspective shifts the focus from merely conserving resources for human use to protecting the environment for its own sake.

A significant issue in environmental ethics is **anthropocentrism**, the belief that human beings are the most important entities in the universe. Anthropocentrism places human needs and desires at the center of ethical considerations, often at the expense of the environment. This mindset has contributed to widespread environmental degradation, as natural resources have been exploited without regard for the long-term consequences.

In response to anthropocentrism, **ecocentrism** and **biocentrism** have emerged as alternative ethical frameworks. Ecocentrism is the belief that ecosystems as a whole —rather than individual species—deserve moral consideration. This approach emphasizes the interconnectedness of all living things and argues that harming one part of an ecosystem can have negative effects on the entire system. Biocentrism, on the other hand, extends moral consideration to all living beings, regardless of their role in an ecosystem or their utility to humans.

Another central theme in environmental ethics is the idea of **sustainability**. Ethical frameworks that prioritize sustainability argue that current generations have a responsibility to ensure that future generations inherit a planet capable of supporting life. This includes protecting natural resources, preserving biodiversity, and mitigating the effects of climate change. The ethical challenge is finding a balance between development and environmental protection, so that human needs are met without compromising the health of the planet.

Environmental justice is also a key concern within environmental ethics. This concept focuses on the fair distribution of environmental benefits and burdens across different communities. Environmental justice advocates argue that marginalized communities often bear the brunt of environmental problems—such as pollution, deforestation, and climate change—despite contributing the least to these issues. This raises questions about the fairness of environmental policies and the ethical obligation to protect vulnerable populations from harm.

The field of environmental ethics has influenced global environmental policy, leading to the development of laws and regulations aimed at conserving ecosystems and addressing environmental degradation. For example, international agreements like the Paris Agreement on climate change reflect ethical concerns about the responsibility of developed nations to reduce greenhouse gas emissions and support developing countries in transitioning to sustainable energy.

Global Environmental Challenges

Global environmental challenges are interconnected issues that affect ecosystems and human societies around the world. These challenges include climate change, biodiversity loss, deforestation, pollution, and resource depletion. Addressing them requires international cooperation and sustainable development strategies to mitigate environmental damage while ensuring human well-being.

One of the most pressing global environmental challenges is **climate change**. Driven primarily by human activities such as the burning of fossil fuels, deforestation, and industrial agriculture, climate change results from the release of greenhouse gases like carbon dioxide and methane into the atmosphere. These gases trap heat, leading to rising global temperatures. The impacts of climate change are wide-ranging: melting ice caps, rising sea levels, more frequent and intense extreme weather events, and disruptions to agricultural production. As global temperatures rise, ecosystems struggle to adapt, leading to shifts in species distributions, the loss of biodiversity, and increased risk of extinction for many species.

Biodiversity loss is another critical challenge. Habitat destruction, pollution, overfishing, and climate change are all contributing to the rapid decline of species worldwide. The loss of biodiversity threatens ecosystem services that humans rely on, such as pollination, water purification, and carbon sequestration. The extinction of key species can lead to cascading effects in ecosystems, where the disappearance of one species affects many others. For example, the loss of top predators can result in overpopulation of prey species, which in turn may degrade vegetation and alter the structure of ecosystems.

Deforestation is a significant driver of both climate change and biodiversity loss. Forests are vital carbon sinks, absorbing large amounts of carbon dioxide from the atmosphere. However, deforestation—often for agriculture, logging, and urban expansion—releases stored carbon, contributing to global warming. It also destroys habitats, leading to the displacement of wildlife and the disruption of ecosystems. Tropical rainforests, such as the Amazon, are particularly vulnerable, and their loss has global consequences, including reduced carbon sequestration and increased climate instability.

Pollution is another global challenge that affects air, water, and soil quality. Air pollution, largely from industrial activities and vehicle emissions, contributes to respiratory diseases and other health problems. It also exacerbates climate change by increasing the concentration of greenhouse gases. Water pollution, from sources like agricultural runoff, industrial waste, and untreated sewage, threatens aquatic ecosystems and contaminates drinking water supplies. Soil pollution, often due to pesticide use and industrial waste disposal, reduces soil fertility and harms plant and animal life.

Resource depletion is a growing concern as the global population increases and consumption patterns rise. Non-renewable resources, such as fossil fuels, minerals, and freshwater, are being used at unsustainable rates. Overextraction of these resources can lead to environmental degradation, such as habitat destruction, soil erosion, and water scarcity. Even renewable resources, like forests and fisheries, are under pressure due to overexploitation. Unsustainable practices in industries like logging, fishing, and agriculture deplete these resources faster than they can regenerate, leading to long-term damage to ecosystems and a reduction in biodiversity.

Water scarcity is a significant component of resource depletion, particularly in regions where freshwater resources are limited or where water management practices are poor. Overuse of water for agriculture, industry, and urban development strains rivers, lakes, and aquifers. Climate change exacerbates this by altering precipitation patterns, leading to droughts in some regions and floods in others. Water scarcity not only threatens human populations but also affects ecosystems that rely on freshwater to function.

Global environmental challenges are compounded by **population growth and urbanization**. As the world's population continues to increase, more land is cleared for housing, infrastructure, and agriculture. This leads to habitat loss, increased energy consumption, and greater waste production. Urban areas also tend to concentrate pollution, and the demand for resources rises with the growing number of people living in cities.

Addressing these global challenges requires **international cooperation** and the implementation of sustainable development goals. Policies aimed at reducing greenhouse gas emissions, protecting endangered species, and conserving natural resources are critical. Multilateral agreements like the Paris Agreement on climate change, international biodiversity conventions, and efforts to combat deforestation are examples of the global response to these issues.

Technological innovation also has a key role in addressing environmental challenges. Advancements in renewable energy, such as solar and wind power, offer alternatives to fossil fuels, while sustainable agriculture practices can help reduce the environmental impact of food production. Conservation efforts, such as reforestation projects and the establishment of marine protected areas, are critical in restoring ecosystems and preserving biodiversity.

However, the success of these efforts depends on the collective will of governments, businesses, and individuals to make sustainable choices. Global environmental challenges are not confined by borders, and solutions must be implemented on a global scale to ensure that both natural ecosystems and human societies can thrive in the future.

CHAPTER 2: ECOSYSTEMS AND ENERGY FLOW

Components of Ecosystems: Biotic and Abiotic Factors

An ecosystem is a community of living organisms interacting with their physical environment. To understand ecosystems fully, we need to explore both their **biotic** and **abiotic** components. These two sets of factors work together to support life, regulate processes like energy flow, and maintain balance within the system.

Biotic factors are the living components of an ecosystem. These include all organisms, from the tiniest microorganisms like bacteria and fungi to large mammals, trees, and humans. Biotic factors can be broken down into three main categories based on their roles: producers, consumers, and decomposers.

Producers, also known as autotrophs, form the base of every ecosystem. They create their own food, usually through photosynthesis, using energy from sunlight, carbon dioxide from the atmosphere, and water from the soil. In terrestrial ecosystems, plants are the primary producers. In aquatic ecosystems, algae and phytoplankton perform this role. These producers are vital because they convert solar energy into a form that can be used by other organisms in the ecosystem. Without producers, ecosystems would lack a primary energy source, and the entire food chain would collapse.

Consumers, or heterotrophs, are organisms that cannot produce their own food. They depend on other organisms for energy and nutrients. Consumers can be classified into different levels, including herbivores, carnivores, and omnivores, depending on their diet.

- **Herbivores**, such as deer and rabbits, feed on plants, making them primary consumers.
- **Carnivores**, like lions and wolves, eat other animals. They are secondary or tertiary consumers, depending on their position in the food chain.
- **Omnivores**, like humans and bears, consume both plants and animals, allowing them to operate at multiple levels in the food chain.

Each type of consumer contributes to energy flow by transferring energy from one level of the ecosystem to another. For example, when a herbivore eats a plant, it takes in the energy the plant has stored from photosynthesis. That energy then moves up the food chain when a carnivore consumes the herbivore.

Decomposers are another vital group of biotic factors. These organisms, which include fungi, bacteria, and certain insects, break down dead plant and animal matter. By doing so, they recycle nutrients back into the ecosystem, making them available for producers to use again. Without decomposers, ecosystems would accumulate dead material, and essential nutrients would become locked away, unavailable for reuse.

On the other side of the ecosystem equation, **abiotic factors** are the non-living components. These include physical and chemical elements like sunlight, water, air, temperature, and soil. While these factors may not be alive, they directly affect how ecosystems function and the kinds of life they support.

Sunlight is arguably the most important abiotic factor because it drives photosynthesis, the process by which producers generate food. Without sunlight, ecosystems would not have the energy needed to support life. The amount of sunlight an ecosystem receives can vary widely depending on its location. For example, tropical rainforests near the equator receive abundant sunlight, allowing for rich biodiversity, while polar ecosystems receive limited sunlight, especially during the winter months, resulting in fewer species and a slower energy flow.

Water is another critical abiotic factor. All living organisms require water for survival. Water availability affects the types of species that can live in an ecosystem. Deserts, where water is scarce, support fewer species compared to wetlands or tropical forests. In addition to its role in sustaining life, water is essential for many biochemical processes and is a key player in energy transfer through ecosystems.

Air is essential for respiration in animals and photosynthesis in plants. Oxygen from the air is used by organisms to release the energy stored in food through cellular respiration. Meanwhile, plants take in carbon dioxide from the atmosphere to produce glucose through photosynthesis. The composition of the air in an ecosystem can also influence its health. For example, an increase in air pollutants like sulfur dioxide can lead to acid rain, which harms plants and aquatic life.

Temperature is another abiotic factor that influences ecosystems. Temperature affects metabolic rates in organisms, determining how quickly they grow, reproduce, and process nutrients. Cold environments, such as the Arctic, have species adapted to slower metabolic processes, while warmer climates like the tropics support faster biological activity and greater biodiversity. Fluctuations in temperature can also affect migration patterns, hibernation, and breeding seasons for various species.

Soil composition affects plant growth, which in turn influences the entire ecosystem. Different types of soil contain varying amounts of nutrients, minerals, and organic matter. For example, nutrient-rich soil can support a diverse range of plant species, which then supports a larger variety of herbivores and, consequently, carnivores. Soil also influences water retention, which affects the availability of water to plants and other organisms.

Trophic Levels and Food Chains/Webs

In an ecosystem, organisms are classified into different trophic levels based on their role in the flow of energy. These levels are organized in a hierarchical structure, with each level representing a step in the transfer of energy through food consumption. Understanding trophic levels helps clarify how energy moves within ecosystems and highlights the interdependence of organisms.

At the foundation of every food chain or food web are the **producers**, also known as **autotrophs**. These organisms, primarily plants, algae, and some bacteria, make their own food through photosynthesis or chemosynthesis. They form the **first trophic level** and are responsible for converting solar energy into chemical energy, which can then be used by other organisms.

Above the producers are the **primary consumers**, or **herbivores**, which feed directly on plants. These organisms form the **second trophic level**. Herbivores, such as deer, rabbits, and insects, consume the energy stored in plants. By eating plants, they transfer energy up the food chain.

Next are the **secondary consumers**, or **carnivores** that feed on herbivores. These organisms occupy the **third trophic level**. Carnivores like wolves, snakes, and birds of prey depend on herbivores for their energy. Some secondary consumers are also **omnivores**, eating both plants and animals, which allows them to access energy from multiple sources within the ecosystem.

Above secondary consumers are **tertiary consumers**, which are often top predators in an ecosystem. These organisms occupy the **fourth trophic level**. Tertiary consumers feed on both secondary consumers and primary consumers. Examples include sharks, eagles, and large cats like lions and tigers. These predators are often at the top of the food chain, meaning they have no natural predators of their own within their ecosystem.

In some ecosystems, there are even **quaternary consumers**, or **apex predators**, which sit at the very top of the food chain. Apex predators, like orcas and polar bears, do not have any natural enemies and are important in maintaining balance by regulating the populations of other species.

The distinction between **food chains** and **food webs** is important. A **food chain** is a linear sequence showing how energy moves from one trophic level to the next. It typically follows a single path, like grass (producer) → rabbit (primary consumer) → fox (secondary consumer). However, ecosystems are rarely this simple. Organisms often have multiple food sources, and a single predator may consume a variety of prey. This complexity is represented in **food webs**, which are more accurate models of energy flow within ecosystems. A food web is a network of interconnected food chains, showing how different organisms feed on multiple species across trophic levels.

In a food web, the removal or decline of a species can have cascading effects throughout the system. For example, if a predator is removed, its prey population may grow unchecked, leading to overgrazing and degradation of plant life. This demonstrates the importance of each organism within the web, highlighting the interconnectedness of ecosystems.

Energy Flow: The 10% Rule and Energy Loss

Energy flow in an ecosystem follows a simple but important principle: energy moves from one trophic level to the next but diminishes as it travels up the food chain. This loss of energy is explained by the **10% rule**, which states that, on average, only 10% of the energy from one trophic level is passed on to the next. The rest of the energy is lost through various processes, such as heat loss, metabolic functions, and waste. Understanding this energy loss is crucial for grasping how ecosystems maintain balance and why top predators are often fewer in number.

Energy enters an ecosystem through **producers**, mainly plants and other photosynthetic organisms. These producers capture sunlight and convert it into chemical energy through photosynthesis. This energy is stored in the form of glucose, which they use for growth, reproduction, and other metabolic processes. However, even at this level, some energy is lost. Plants use energy for respiration, and a portion is lost as heat. Thus, the energy stored in plants is less than the total energy they capture from sunlight.

When a **primary consumer** (herbivore) eats a plant, it only gains a fraction of the plant's energy. According to the 10% rule, if a plant stores 1,000 units of energy, the herbivore only receives about 100 units from consuming it. The remaining 900 units are lost. This energy loss occurs because the plant uses a significant portion of its energy for respiration, maintenance, and reproduction, leaving less energy available for the consumer. Additionally, not all parts of the plant are consumed or digested, so some energy is lost in the form of waste.

Similarly, when a **secondary consumer** (carnivore) eats the herbivore, it only acquires about 10% of the energy that the herbivore had. If the herbivore had 100 units of energy, the carnivore receives just 10 units. The carnivore also loses energy through metabolic processes like movement, digestion, and heat production. As energy moves up each trophic level, less and less is available, which limits the number of organisms that can exist at higher levels.

At the **tertiary consumer** level, the energy transferred is even smaller. If a secondary consumer had 10 units of energy, a tertiary consumer only receives about 1 unit. This dramatic reduction in available energy explains why ecosystems have far fewer tertiary consumers and apex predators compared to primary consumers. Energy is simply insufficient to support large populations at the top of the food chain.

This **energy loss** at each level also explains why food chains rarely extend beyond four or five trophic levels. Beyond that, the energy available is too minimal to sustain additional levels. In some ecosystems, like deep-sea environments or regions with limited resources, food chains may only have two or three levels due to the scarcity of energy.

The 10% rule is not absolute—some ecosystems might transfer slightly more or less energy depending on factors like species efficiency and environmental conditions. For example, cold-blooded animals like reptiles may be more energy-efficient than warm-blooded mammals because they don't need to use energy to maintain a constant body temperature. However, the principle of significant energy loss remains consistent across all ecosystems.

Another critical concept tied to energy loss is **energy efficiency**. Organisms that are lower on the food chain, such as plants or herbivores, make more efficient use of available energy. This is why ecosystems with more plant biomass and fewer top predators can support larger populations. In agricultural systems, for instance, growing plants directly for human consumption is more energy-efficient than feeding plants to livestock and then consuming meat.

Keystone Species and Ecosystem Balance

A keystone species is one whose presence and role in an ecosystem have a disproportionately large impact on maintaining the balance and structure of that system. These species play a critical part in the overall functioning of ecosystems, influencing the diversity and abundance of other species, as well as ecological processes like predation, competition, and nutrient cycling. Without keystone species, ecosystems can undergo dramatic changes, often leading to a collapse or significant shift in the ecosystem structure.

One of the most well-known examples of a keystone species is the **gray wolf** in Yellowstone National Park. Prior to the reintroduction of wolves in 1995, the elk population in Yellowstone had grown unchecked, as there were no natural predators to control it. This led to overgrazing of vegetation, particularly willows and aspens, which in turn affected other species like beavers and songbirds. With the return of wolves, the elk population was reduced, which allowed vegetation to recover. The increase in plant life led to healthier streams, providing habitat for fish and beavers, and benefitting a range of species throughout the ecosystem. In this case, wolves, as top predators, indirectly shaped the environment by controlling prey populations, highlighting their role as a keystone species.

Keystone species aren't always predators. In some ecosystems, **herbivores** or even **plants** can serve as keystone species. For example, in African savannas, elephants are considered a keystone species due to their role in shaping the landscape. Elephants knock down trees, allowing grasses to thrive, which supports herbivores like zebras and antelope. Without elephants, the landscape would become more forested, reducing the grassland and altering the entire ecosystem. This shows how a single species can influence both the physical environment and the variety of species within it.

Pollinators such as bees can also act as keystone species, particularly in ecosystems where plants rely on animals to reproduce. Bees pollinate a wide variety of plants, including many species that provide food and habitat for other organisms. If bee populations decline, the plants they pollinate may also decline, leading to reduced food sources for herbivores and a potential cascading effect throughout the ecosystem.

In marine ecosystems, **sea otters** are a keystone species that help maintain the balance of kelp forests. Sea otters feed on sea urchins, which are herbivores that graze on kelp. Without otters to keep the sea urchin population in check, the urchins can overgraze and destroy entire kelp forests. Kelp forests provide important habitat and food for a wide range of marine species, so the presence of otters is essential for the health of this ecosystem.

The loss of a keystone species can cause a **trophic cascade**, a chain reaction that disrupts multiple levels of the ecosystem. When a keystone species is removed, prey species may grow uncontrollably, leading to overconsumption of resources. This imbalance can reduce biodiversity and degrade ecosystem services, such as water purification, pollination, and soil fertility. In many cases, the absence of a keystone species leads to the dominance of a few species, which outcompete others and reduce overall ecosystem diversity.

The importance of keystone species is not just limited to their direct interactions with other organisms. They also have a critical role in **maintaining ecosystem functions** like nutrient cycling and habitat structure. For example, in tropical rainforests, certain large fruit-eating animals, such as primates or large birds, act as keystone species by dispersing seeds. These animals help maintain forest diversity by spreading seeds over wide areas, ensuring the regeneration of different plant species. Without them, many plants would fail to reproduce and the structure of the forest would change, affecting countless other species.

In short, keystone species are fundamental to the health and stability of ecosystems. Their actions, whether through predation, competition, or habitat modification, maintain the balance that allows ecosystems to function. Protecting keystone species is crucial for conservation efforts because their loss can lead to the collapse of ecosystems and the loss of biodiversity.

Ecosystem Resilience and Recovery After Disturbance

Ecosystem resilience refers to the ability of an ecosystem to absorb disturbances, adapt to changing conditions, and recover without losing its essential structure and functions. Disturbances can come in many forms—natural events like wildfires, floods, or storms, as well as human-caused events like deforestation, pollution, and climate change. A resilient ecosystem can bounce back after these disturbances, while less resilient systems may suffer long-term damage or even collapse.

The **degree of resilience** in an ecosystem depends on several factors, including biodiversity, the complexity of interactions between species, and the health of ecological processes like nutrient cycling and energy flow. **Biodiversity** is particularly important because ecosystems with a wide variety of species are often more capable of responding to disturbances. When multiple species fulfill similar roles in an ecosystem, the loss of one species can be compensated for by another. For example, if a drought reduces the population of one plant species, other drought-resistant plants may fill the gap, ensuring that herbivores still have food sources. This redundancy contributes to the stability and resilience of ecosystems.

Another key factor in ecosystem resilience is the presence of **keystone species**, which often are important in helping ecosystems recover from disturbances. Keystone species regulate populations of other species, maintain balance in food webs, and support ecosystem processes. If a keystone species is removed, an ecosystem's resilience may be weakened, making it more vulnerable to further disturbances.

Natural disturbances, such as **wildfires**, can actually enhance ecosystem resilience in some cases. For example, fire is a natural part of many ecosystems, such as grasslands and some forests, where it has a role in nutrient cycling and promoting new growth. In these ecosystems, many species are adapted to fire, and periodic burns can help maintain biodiversity by clearing out dead material, controlling insect pests, and creating space for new plants to grow. However, the frequency and intensity of fires are crucial—too frequent or severe fires can overwhelm an ecosystem's ability to recover.

Human-caused disturbances, on the other hand, often reduce resilience by degrading the natural processes that support recovery. **Deforestation**, for instance, removes trees that help regulate water cycles, prevent soil erosion, and support biodiversity. Without these trees, the ecosystem's ability to recover from other disturbances, like floods or droughts, is significantly reduced. Similarly, **pollution**

23

can weaken ecosystems by contaminating soil, water, and air, making it harder for plants and animals to survive and reproduce.

Climate change is one of the most significant threats to ecosystem resilience today. Rising temperatures, shifting precipitation patterns, and increasing frequency of extreme weather events can push ecosystems beyond their thresholds, making it harder for them to recover from disturbances. For example, coral reefs are highly sensitive to temperature changes, and even small increases in water temperature can lead to coral bleaching, where corals expel the algae that live in their tissues and provide them with nutrients. If water temperatures remain high for extended periods, the corals cannot recover, leading to widespread reef degradation.

Ecosystems with high **connectivity** tend to be more resilient. Connectivity refers to the way species and resources move across different parts of the ecosystem. For example, rivers connect aquatic and terrestrial ecosystems, transporting nutrients and organisms between them. When ecosystems are fragmented, such as by roads, dams, or urban development, connectivity is disrupted, making it harder for species to migrate, reproduce, or find food after a disturbance. Fragmentation can also prevent ecosystems from redistributing resources, like water or nutrients, which are essential for recovery.

Despite the challenges posed by disturbances, ecosystems have remarkable capacity for **recovery**. After a disturbance, ecosystems often go through a process called **ecological succession**, where species gradually repopulate the area. **Primary succession** occurs in environments where no previous ecosystem existed, such as after a volcanic eruption or glacial retreat. In these cases, pioneer species, like lichens or grasses, are the first to colonize the barren landscape, creating soil and conditions that allow other species to establish themselves. **Secondary succession** happens in areas where an ecosystem has been disturbed but still retains some soil and life, such as after a wildfire or human agricultural abandonment. In secondary succession, ecosystems tend to recover faster because some components of the previous ecosystem remain intact.

The ability of an ecosystem to return to its previous state depends on the **intensity of the disturbance**, the **availability of resources**, and the **time** allowed for recovery. Some ecosystems can recover within a few years, while others may take decades or centuries. Forests, for example, may take decades to regrow after clear-cutting, while wetlands can sometimes recover more quickly if water flow and native species are restored.

CHAPTER 3: BIOGEOCHEMICAL CYCLES

The Carbon Cycle

The carbon cycle is one of Earth's most essential biogeochemical cycles, moving carbon through the atmosphere, oceans, soil, and living organisms. It's central to life on Earth and climate regulation. Carbon is found in various forms: carbon dioxide (CO_2) in the atmosphere, dissolved carbon in oceans, and organic matter in soils and living organisms. Understanding how carbon moves through these reservoirs gives insight into global climate patterns, energy flow, and ecosystem dynamics.

The carbon cycle begins with the atmosphere, where carbon exists mostly as carbon dioxide (CO_2). This CO_2 is absorbed by plants through **photosynthesis**, the process by which plants convert sunlight, water, and carbon dioxide into glucose (a form of sugar) and oxygen. Plants use this glucose to build their tissues and grow. During this process, they remove CO_2 from the air, temporarily storing it as organic carbon. **Phytoplankton**, tiny plants in the ocean, also perform photosynthesis, making them key players in the global carbon cycle. Oceans act as both a carbon sink, absorbing CO_2 from the atmosphere, and as a carbon source, releasing CO_2 back into the atmosphere.

As plants grow and animals consume them, carbon moves through the **food chain**. Herbivores, such as deer and rabbits, eat plants and obtain carbon from plant tissues. Carnivores, like wolves, then eat herbivores, passing carbon up the food chain. When these animals breathe, they release carbon dioxide back into the atmosphere through **respiration**. This is a critical aspect of the carbon cycle—living organisms continuously exchange carbon with the atmosphere as they respire.

Another significant process in the carbon cycle is **decomposition**. When plants and animals die, decomposers like bacteria and fungi break down their organic matter. During this process, the carbon in their bodies is converted back into CO_2 or methane (CH_4) and released into the atmosphere. This recycling of carbon ensures that the nutrient cycles through the ecosystem, supporting new growth.

The oceans store vast amounts of carbon, not just in the water itself but also in marine organisms and sediments. When marine organisms, like plankton and shellfish, die, their carbon-rich bodies sink to the ocean floor. Over time, this organic matter can become buried under layers of sediment, forming **sedimentary rock** or even fossil fuels. In this way, carbon is stored in the ocean's depths for long periods. Some of the carbon that reaches the deep ocean is returned to the surface through **upwelling**, a process in which deep, nutrient-rich waters rise to the surface, bringing carbon with them. This can then be released back into the atmosphere, continuing the cycle.

Fossil fuels, including coal, oil, and natural gas, are another important part of the carbon cycle. These fuels were formed millions of years ago from the remains of ancient plants and animals that were buried and subjected to intense heat and pressure. Over geological time, these organic materials were transformed into energy-rich carbon stores. When humans burn fossil fuels for energy, they release large amounts of carbon dioxide into the atmosphere, a process known as **combustion**. This combustion of fossil fuels is a major source of CO_2 emissions, contributing to the enhanced greenhouse effect and global warming.

The ocean also interacts with the atmosphere in terms of **carbon absorption and release**. Oceans absorb CO_2 from the atmosphere, where it dissolves in the water. Some of this dissolved carbon reacts with water to form carbonic acid (H_2CO_3), which then breaks down into bicarbonate (HCO_3^-) and carbonate ions (CO_3^{2-}). This process helps regulate the amount of CO_2 in the atmosphere, but as more CO_2 is absorbed, the oceans become more acidic—a phenomenon known as **ocean acidification**. This increased acidity can have harmful effects on marine life, particularly organisms with calcium carbonate shells, like corals and shellfish, as it weakens their ability to form shells.

In addition to these natural processes, **human activities** have significantly altered the carbon cycle. The burning of fossil fuels for energy, industrial processes, and deforestation have increased the amount of carbon dioxide in the atmosphere. Deforestation, in particular, is problematic because trees act as carbon sinks—they absorb CO_2 during photosynthesis and store it as biomass. When trees are cut down or burned, this stored carbon is released back into the atmosphere, contributing to elevated levels of greenhouse gases. Land-use changes like agriculture and urbanization also reduce the Earth's ability to absorb and store carbon.

Carbon sequestration is a process by which carbon is stored long-term in forests, soils, oceans, and underground geological formations. Forests and other vegetative landscapes are crucial carbon sinks, as they absorb more carbon than they release. Similarly, practices like **no-till farming** and **reforestation** help enhance carbon storage in soils and plants. Technological methods, such as **carbon capture and storage (CCS)**, are being developed to capture CO_2 emissions from industrial processes and store them underground in geological formations, preventing them from entering the atmosphere.

Volcanic activity also influences the carbon cycle. **Volcanoes** release carbon dioxide stored deep within the Earth into the atmosphere during eruptions. While volcanic CO_2 emissions are relatively small compared to human-caused emissions, they are a natural part of the carbon cycle and have influenced Earth's climate over long geological time periods.

Another long-term aspect of the carbon cycle involves the **rock cycle**. Carbon stored in the Earth's crust as carbonate rocks (like limestone) can be released through **weathering** processes. When rainwater, slightly acidic due to dissolved CO_2, falls on these rocks, it reacts with the carbonates, releasing carbon back into the atmosphere or into rivers and oceans. This process, though slow, is vital in regulating atmospheric CO_2 over geological timescales.

The **carbon cycle** is a dynamic and interconnected system. It links the atmosphere, land, oceans, and living organisms in a continuous exchange of carbon. As carbon moves through the cycle, it influences everything from climate patterns to the productivity of ecosystems. **Human activities**, particularly the burning of fossil fuels, have upset the balance of the carbon cycle, contributing to the accumulation of greenhouse gases and the warming of the planet. Understanding how carbon moves through the environment is essential for addressing climate change and managing the Earth's carbon resources.

The Nitrogen Cycle

The nitrogen cycle is a vital process that circulates nitrogen through the atmosphere, soil, and living organisms. Nitrogen is a key component of amino acids, proteins, and DNA, making it essential for all life forms. However, despite nitrogen's abundance in the atmosphere—around 78%—it must undergo several transformations to be usable by plants and animals. This transformation is what the nitrogen cycle manages.

Nitrogen in the atmosphere exists primarily as nitrogen gas (N_2), which is highly stable and cannot be directly used by most living organisms. The first step in the nitrogen cycle is **nitrogen fixation**, where specialized bacteria in the soil and water, such as those in the roots of legumes, convert atmospheric nitrogen into ammonia (NH_3). These nitrogen-fixing bacteria use enzymes to break the strong triple bond between nitrogen atoms, allowing nitrogen to combine with hydrogen and form ammonia. This process makes nitrogen available to plants, which absorb it through their roots.

Once ammonia is available in the soil, **nitrification** occurs. This is a two-step process carried out by nitrifying bacteria. First, bacteria called **Nitrosomonas** convert ammonia (NH_3) into nitrite (NO_2^-), a form of nitrogen that is still toxic to plants. Then, another group of bacteria, **Nitrobacter**, convert nitrite into nitrate (NO_3^-), a form of nitrogen that plants can safely absorb and use for growth.

Plants take up nitrates from the soil and use them to build proteins and other essential compounds. When animals eat these plants, they, too, take in nitrogen. This step is known as **assimilation**. The nitrogen absorbed by plants and animals becomes part of their tissues, supporting various biological functions like growth, reproduction, and cell repair.

When plants and animals die, or when animals excrete waste, nitrogen returns to the soil in the form of organic matter. **Decomposers** such as bacteria and fungi break down this matter through a process called **ammonification**, releasing ammonia back into the soil. This ammonia can then be used again by plants or undergo nitrification, continuing the cycle.

Not all nitrogen stays in the soil. Some is lost through a process known as **denitrification**, where denitrifying bacteria convert nitrates (NO_3^-) back into nitrogen gas (N_2) or nitrous oxide (N_2O), both of which are released into the atmosphere. This step closes the nitrogen cycle, returning nitrogen to its gaseous state and completing the loop.

Human activities have significantly altered the nitrogen cycle, particularly through the use of synthetic fertilizers. These fertilizers contain large amounts of nitrogen, which can improve crop yields but often lead to excessive nitrogen runoff into water bodies. This runoff can cause **eutrophication**, where nitrogen-rich waters promote the overgrowth of algae. As algae die and decompose, oxygen in the water is depleted, leading to dead zones that cannot support aquatic life.

Industrial nitrogen fixation, primarily through the Haber-Bosch process, also adds significant amounts of nitrogen to ecosystems. This process converts atmospheric nitrogen into ammonia on a massive scale for fertilizer production, further disrupting natural nitrogen cycles. Additionally, the burning of fossil fuels releases nitrogen oxides (NO_x) into the atmosphere, contributing to smog and acid rain, which negatively impact ecosystems and human health.

The Water Cycle (Hydrological Cycle)

The water cycle, also known as the hydrological cycle, is the continuous movement of water on, above, and below the Earth's surface. Water circulates between the atmosphere, land, and oceans, constantly changing its state between liquid, vapor, and ice. This process is essential for regulating climate, supporting life, and shaping landscapes.

The cycle begins with **evaporation**, the process where water from oceans, lakes, rivers, and even plants turns into water vapor due to the heat from the Sun. Oceans, being the largest bodies of water, contribute the most to evaporation. However, water can also evaporate from soil and through **transpiration**, where plants release water vapor from their leaves. Together, evaporation and transpiration are often referred to as **evapotranspiration**. This movement of water vapor into the atmosphere is crucial for forming clouds and precipitation.

As water vapor rises into the atmosphere, it cools and undergoes **condensation**. During this phase, the water vapor turns back into liquid droplets or ice crystals, forming clouds. Condensation is responsible for cloud formation, and the type of cloud that forms depends on factors like temperature, humidity, and altitude. For instance, when warm, moist air rises and cools rapidly, large, puffy clouds called cumulus clouds may form. If the air cools more slowly, stratus clouds, which are flatter and more spread out, will form.

The next stage in the water cycle is **precipitation**. When water droplets or ice crystals in clouds become too heavy, they fall back to Earth as rain, snow, sleet, or hail. Precipitation is the main way water returns from the atmosphere to the Earth's surface. The form of precipitation depends on atmospheric conditions. If temperatures are above freezing, rain is the most likely outcome, but when temperatures are colder, snow or ice forms.

Once water reaches the ground, it can follow several paths. Some of it will **infiltrate** the soil, replenishing groundwater supplies in a process called **percolation**. Groundwater is stored in aquifers, which provide water for drinking, agriculture, and other uses. **Runoff** occurs when water flows over the land's surface into streams, rivers, and lakes, eventually making its way back to the ocean. Runoff is essential for moving nutrients, sediments, and dissolved substances across ecosystems, but excessive runoff can lead to flooding, especially when the ground is saturated or impermeable.

Water stored in the ground as **groundwater** may take years, decades, or even centuries to re-enter the cycle, depending on how deep it is. In contrast, water in rivers and lakes moves more quickly through the cycle. Groundwater is an essential source of fresh water for human consumption, irrigation, and industrial use. It also supports ecosystems by providing moisture for plants and feeding into rivers and lakes.

Another pathway is **sublimation**, a process in which ice and snow transform directly into water vapor without first melting into liquid. This is more common in colder climates, particularly in high-altitude regions or during colder seasons, where snowfields and glaciers gradually lose mass through sublimation.

Water that reaches oceans, lakes, or rivers eventually **evaporates** again, completing the cycle. However, water can be temporarily stored in glaciers, ice caps, and snowfields for long periods, creating **ice reserves** that are important in regulating Earth's climate. The melting of these ice reserves due to global warming is currently increasing the amount of liquid water in the oceans, contributing to rising sea levels.

The water cycle is also influenced by **human activities**. Dams, reservoirs, and water extraction for agriculture or industry can alter natural water flow. Urbanization increases runoff by covering land with impermeable surfaces like concrete and asphalt, reducing infiltration and increasing the risk of flooding. Pollution, such as chemicals or waste products, can enter the water cycle through runoff, contaminating rivers, lakes, and even groundwater, which affects ecosystems and human health.

Climate change is impacting the water cycle by altering precipitation patterns and increasing evaporation rates. Warmer temperatures lead to more intense and frequent storms, altering rainfall distribution and causing droughts in some regions and floods in others. Rising temperatures also accelerate the melting of glaciers and polar ice caps, which contributes to sea level rise and changes in freshwater availability.

The water cycle is a complex, self-sustaining system that not only supports all life forms but also shapes Earth's weather, climate, and physical landscapes. It connects all major Earth systems, making it one of the most important processes for maintaining balance within the biosphere.

Human Impact on Biogeochemical Cycles

Human activities have significantly altered the natural biogeochemical cycles, disrupting the balance of essential elements like carbon, nitrogen, phosphorus, and water in ecosystems. These changes have far-reaching consequences for biodiversity, ecosystem stability, and climate regulation.

One of the most profound human impacts is on the **carbon cycle**. The burning of fossil fuels, such as coal, oil, and natural gas, for energy releases large amounts of carbon dioxide (CO_2) into the atmosphere. This has led to increased atmospheric CO_2 concentrations, contributing to the greenhouse effect and global warming. Deforestation further exacerbates the problem by reducing the number of trees that can absorb CO_2 through photosynthesis. The result is a buildup of carbon in the atmosphere, which intensifies climate change, leading to more frequent extreme weather events, rising sea levels, and shifts in ecosystems.

The **nitrogen cycle** has also been significantly affected by human activities, particularly through the use of synthetic fertilizers in agriculture. Nitrogen fertilizers, which are rich in reactive nitrogen compounds like ammonia and nitrates, are applied to increase crop yields. However, much of this nitrogen is not absorbed by plants and ends up leaching into rivers, lakes, and groundwater. This leads to **eutrophication**, where excess nutrients promote algae growth in aquatic systems. As algae die and decompose, oxygen is depleted, creating dead zones that cannot support fish or other aquatic life. Additionally, the release of nitrogen oxides (NO_x) from fossil fuel combustion contributes to air pollution and acid rain, which harms ecosystems by acidifying soils and water bodies.

The **phosphorus cycle** has been disrupted mainly by agricultural runoff and wastewater discharge. Phosphorus, which is used in fertilizers and detergents, accumulates in water bodies, contributing to eutrophication similar to nitrogen. Unlike nitrogen, phosphorus is a limiting nutrient in many ecosystems, so even small increases can have dramatic effects. Excess phosphorus promotes the overgrowth of algae and aquatic plants, leading to oxygen depletion, fish kills, and the loss of biodiversity. Mining for phosphate rock to produce fertilizers also causes habitat destruction and pollution in mining areas.

Human activity has altered the **water cycle** (or hydrological cycle) through deforestation, urbanization, and the construction of dams and reservoirs. Deforestation reduces the ability of forests to absorb and retain water, leading to increased surface runoff and soil erosion. Urbanization covers land with impermeable surfaces like asphalt and concrete, which prevents water from infiltrating the soil and increases the risk of flooding. The diversion of water for agriculture, industry, and domestic use has depleted many freshwater resources, including rivers, lakes, and aquifers, leading to water scarcity in some regions. In addition, climate change is intensifying the water cycle, with more extreme droughts, floods, and changes in precipitation patterns.

In the **sulfur cycle**, human activities like burning fossil fuels and refining oil have increased sulfur emissions, primarily in the form of sulfur dioxide (SO_2). These emissions contribute to **acid rain**, which harms forests, aquatic ecosystems, and soil fertility by lowering pH levels. Acid rain can also erode buildings and infrastructure, causing economic damage.

Humans have disrupted the natural flow of these biogeochemical cycles in ways that exceed the planet's capacity to absorb or adjust. The consequences of these disruptions include the loss of biodiversity, the degradation of ecosystems, and worsening climate conditions. Rebalancing these cycles requires reducing human impacts, such as lowering carbon emissions, promoting sustainable agriculture, and improving waste management practices.

Sulfur Cycle and Its Environmental Effects

The sulfur cycle describes the movement of sulfur through the environment, including the atmosphere, lithosphere (Earth's crust), hydrosphere (oceans and water bodies), and biosphere (living organisms). Sulfur is an essential element for life, having a critical role in proteins and enzymes. However, human activities have significantly altered the sulfur cycle, leading to environmental issues like acid rain and the degradation of ecosystems.

Sulfur naturally occurs in several forms, including **sulfates (SO_4^{2-})**, **sulfur dioxide (SO_2)**, and **hydrogen sulfide (H_2S)**. Most sulfur in the environment is stored in rocks and minerals. When volcanic activity or weathering breaks down these rocks, sulfur is released into the atmosphere or into bodies of water. In the atmosphere, sulfur dioxide (SO_2) can be emitted by volcanic eruptions or by the decay of organic matter in low-oxygen environments, such as wetlands or swamps.

Sulfur in the atmosphere undergoes several transformations. **Sulfate aerosols**, which form from sulfur compounds, can reflect sunlight, influencing the Earth's climate by cooling the atmosphere. Sulfur dioxide (SO_2), a gas, reacts with water vapor to form sulfuric acid (H_2SO_4), a key component of **acid rain**. This transformation begins when sulfur dioxide is oxidized in the atmosphere, turning into sulfur trioxide (SO_3), which then combines with water molecules to produce sulfuric acid. Acid rain can fall far from the original source of sulfur emissions, affecting ecosystems and infrastructure across vast regions.

Acid rain has serious environmental effects. When sulfuric acid in rainwater reaches the soil, it can **acidify** the soil, lowering its pH and leaching essential nutrients like calcium and magnesium, which are vital for plant growth. This can reduce agricultural productivity and harm forest ecosystems. Acidic soils can also release toxic metals like aluminum, which can be harmful to plants and microorganisms.

In aquatic environments, acid rain can lower the pH of lakes, rivers, and streams, making the water too acidic for fish and other aquatic life. Acidified water can kill sensitive species, disrupt reproduction, and reduce biodiversity. For example, fish species such as trout and salmon are particularly vulnerable to changes in pH, and acidification can lead to their decline or disappearance in affected areas.

The sulfur cycle is closely linked with human activities, especially the burning of fossil fuels. Coal, oil, and natural gas contain sulfur, and when they are burned, sulfur dioxide (SO_2) is released into the atmosphere. Industrial processes, particularly those involved in metal refining, also emit significant amounts of sulfur dioxide. These emissions have been a major contributor to the formation of acid rain, particularly in regions with heavy industrial activity. The widespread use of coal for power generation and oil refining has intensified sulfur emissions, disrupting the natural sulfur cycle.

To mitigate the harmful effects of sulfur emissions, many countries have implemented **sulfur dioxide regulations** aimed at reducing air pollution. Technologies like **flue-gas desulfurization** (often called "scrubbers") are installed in coal-fired power plants to remove sulfur dioxide from exhaust gases before they are released into the atmosphere. These technologies have been effective in reducing sulfur emissions and lowering the incidence of acid rain in many industrialized regions.

Sulfur also moves through the **biosphere**. Plants take up sulfur in the form of sulfates (SO_4^{2-}) from the soil, incorporating it into amino acids and proteins. Animals then obtain sulfur by consuming plants or other animals. When plants and animals die, sulfur is returned to the soil as organic matter decomposes, and sulfur compounds are broken down by bacteria. Some bacteria can also convert sulfur compounds into hydrogen sulfide (H_2S), a gas that can be released back into the atmosphere or used by certain microbes in **chemosynthesis**, especially in deep-sea environments near hydrothermal vents.

In **marine ecosystems**, sulfur is vital. Sulfate ions are abundant in seawater, and sulfur is involved in the formation of **dimethyl sulfide (DMS)**, a compound that influences cloud formation and, consequently, the climate. Phytoplankton, which are tiny plant-like organisms in the ocean, produce DMS as a byproduct of their metabolism. When released into the atmosphere, DMS contributes to cloud condensation processes, indirectly affecting weather patterns and the global climate system.

CHAPTER 4: THE ATMOSPHERE AND CLIMATE SYSTEMS

Layers of the Atmosphere

The atmosphere is a layered blanket of gases surrounding Earth, extending from the surface up into space. These layers are classified based on temperature changes and the physical properties of the gases they contain. Each layer has a different role in regulating Earth's climate, supporting life, and protecting the planet from harmful solar radiation. The atmosphere is divided into five main layers: the troposphere, stratosphere, mesosphere, thermosphere, and exosphere. Each layer has distinct characteristics that influence weather patterns, climate systems, and how energy moves through the atmosphere.

The Troposphere

The **troposphere** is the lowest layer of the atmosphere, extending from Earth's surface to about 8-15 kilometers (5-9 miles) above sea level, depending on location. It's thicker at the equator and thinner at the poles. This layer contains most of the atmosphere's mass—about 75%—and nearly all of its water vapor and clouds. The troposphere is where **weather** occurs, including wind, rain, and storms. This is because water vapor, which is essential for cloud formation and precipitation, is concentrated in this layer.

Temperature decreases with altitude in the troposphere. On average, the temperature drops by about 6.5°C for every kilometer you ascend. This is why mountain tops are colder than sea level. The top of the troposphere, known as the **tropopause**, marks a boundary where the temperature stops decreasing and becomes stable. Above this, the temperature starts to rise again as you enter the next layer, the stratosphere.

The troposphere also contains the air we breathe, composed mostly of nitrogen (78%) and oxygen (21%), with trace amounts of other gases like argon, carbon dioxide, and water vapor. The movement of air in this layer is driven by **convection**, where warm air rises and cool air sinks, creating the atmospheric circulation patterns that drive global weather systems. This movement is influenced by the Earth's rotation, which causes the **Coriolis effect**, deflecting wind patterns and creating the trade winds, westerlies, and polar easterlies.

The Stratosphere

The **stratosphere** sits directly above the troposphere, extending from about 15 kilometers to around 50 kilometers (9 to 31 miles) above the Earth. Unlike the troposphere, the temperature in the stratosphere increases with altitude. This temperature increase is due to the presence of the **ozone layer**, which absorbs the Sun's ultraviolet (UV) radiation, converting it into heat. The ozone layer is crucial for protecting living organisms on Earth from harmful UV rays, which can cause skin cancer and damage ecosystems.

The stratosphere is more stable than the turbulent troposphere below. Air in this layer doesn't mix much vertically, so weather systems do not extend into the stratosphere. This stability makes the stratosphere a prime area for high-altitude aircraft and weather balloons. The **stratopause**, the boundary between the stratosphere and the next layer, the mesosphere, is located around 50 kilometers above the Earth's surface.

The Mesosphere

Above the stratosphere lies the **mesosphere**, which extends from about 50 to 85 kilometers (31 to 53 miles) above Earth. In the mesosphere, temperatures drop once again as you go higher. It is the coldest layer of the atmosphere, with temperatures falling as low as -90°C (-130°F) near the top. This layer is also where most **meteors** burn up upon entering Earth's atmosphere, creating the streaks of light we see as shooting stars. The thin air in the mesosphere generates enough friction to cause meteors to vaporize before they can reach the lower layers of the atmosphere.

The mesosphere is not well understood compared to other layers because it is difficult to study. It's too high for aircraft and weather balloons but too low for most satellites. However, research has shown that **noctilucent clouds**, a rare type of cloud that forms high in the atmosphere, are found in this layer. The **mesopause** marks the upper boundary of the mesosphere and the start of the thermosphere.

The Thermosphere

The **thermosphere** begins around 85 kilometers above the Earth's surface and extends up to about 600 kilometers (373 miles). Temperatures in the thermosphere can soar to 2,500°C (4,500°F) or higher due to the Sun's intense radiation. However, because the air is so thin in this layer, it wouldn't feel hot to a human. The few gas molecules present are spread far apart, so there's not enough density to conduct heat in the way we experience it at the surface.

The thermosphere is where **space begins**, and it is home to the **International Space Station (ISS)**, which orbits within this layer. The thermosphere is also the region where **auroras** occur—bright displays of light seen near the polar regions, known as the aurora borealis in the Northern Hemisphere and the aurora australis in the Southern Hemisphere. These light displays happen when charged particles from the Sun interact with Earth's magnetic field and collide with gases like oxygen and nitrogen in the thermosphere, causing them to glow.

In the thermosphere, the air is ionized, meaning that it is filled with charged particles. This is why this layer is also referred to as the **ionosphere**. The ionosphere is critical for radio communication, as it reflects radio waves back to Earth, enabling long-distance communication. The boundary at the top of the thermosphere is the **thermopause**, beyond which lies the exosphere.

The Exosphere

The **exosphere** is the outermost layer of Earth's atmosphere, extending from around 600 kilometers to 10,000 kilometers (373 to 6,200 miles) above the planet. This layer gradually transitions into the vacuum of space. The air in the exosphere is extremely thin, with gas molecules like hydrogen and helium spaced so far apart that they rarely collide. Because of this, there is no clear boundary between the exosphere and outer space.

In the exosphere, satellites orbit the Earth. The gravitational pull is still present, though very weak, allowing satellites to remain in orbit for extended periods. Spacecraft passing through the exosphere experience almost no atmospheric drag, which makes this region ideal for placing satellites that need to stay in stable orbits.

Though the exosphere contains very few particles, it has a role in **shielding Earth** from cosmic radiation and solar wind. Particles that escape the exosphere drift off into space, but those that remain are part of Earth's extended atmosphere. Due to its low density, the exosphere is where atoms and molecules can escape into space, a process known as **atmospheric escape**.

The Importance of Atmospheric Layers

Each of these layers has a specific function that is critical to maintaining life on Earth and regulating the planet's climate systems. The troposphere is where all weather events take place, driven by temperature differences and air circulation. The stratosphere's ozone layer protects living organisms from harmful UV radiation, while the mesosphere shields the planet from meteors. The thermosphere is important for space exploration, communications, and protecting Earth from solar wind, while the exosphere serves as the boundary between the atmosphere and space.

Together, these layers work to protect Earth, control weather patterns, and support life. Understanding how they interact with one another helps scientists predict climate changes, track weather systems, and explore the connections between Earth's atmosphere and space.

Earth's Climate Zones

Earth's climate zones are categorized based on temperature, precipitation, and general weather patterns. These zones, influenced by factors like latitude, altitude, and proximity to oceans, shape the variety of climates found across the planet. The major climate zones include tropical, temperate, and polar regions, with subzones such as deserts and highland climates offering more localized variation. Each zone has distinct characteristics that affect ecosystems, human settlements, and agricultural practices.

The **tropical climate zone** lies between the Tropic of Cancer and the Tropic of Capricorn, approximately 23.5° north and south of the equator. This region experiences **warm temperatures year-round** due to the direct sunlight it receives. Average monthly temperatures usually exceed 18°C (64°F), and there is minimal variation between seasons. The tropical zone is known for its high humidity and significant rainfall, particularly in rainforests, which can receive more than 2,000 millimeters (79 inches) of rain annually. Tropical rainforests, such as the Amazon and Congo, host some of the most biodiverse ecosystems on Earth. However, there are also **tropical savannas**, which have distinct wet and dry seasons, with less rainfall overall and grasslands dominating the landscape. Tropical zones are often affected by **monsoons**, which are seasonal winds bringing heavy rains that are essential for agriculture in many regions, particularly in South Asia.

Moving away from the equator, you enter the **temperate climate zone**, which lies between approximately 23.5° and 66.5° latitude in both hemispheres. This zone is characterized by **four distinct seasons**—winter, spring, summer, and autumn. The temperature and weather patterns vary significantly between summer and winter, with some regions experiencing cold winters and hot summers. Precipitation in temperate climates can range from moderate to high, depending on the region's proximity to oceans and mountains. **Temperate forests**, such as those found in parts of North America, Europe, and East Asia, are common in these areas. The temperate zone includes both **oceanic climates**, where proximity to large bodies of water moderates temperature swings, and **continental climates**, where temperatures vary more dramatically due to the distance from the sea. Cities like New York and Paris have temperate climates, with cold winters, warm summers, and moderate precipitation throughout the year.

At the poles, the **polar climate zone** covers the areas north of the Arctic Circle and south of the Antarctic Circle, roughly beyond 66.5° latitude. These regions experience **extremely cold temperatures** due to the low angle of sunlight, and in some areas, the Sun does not rise for months during the winter. In the **tundra**, which is found in the polar zones, temperatures are cold year-round, and the soil is often frozen (permafrost). The short summer allows for some plant growth, but the biodiversity is much lower compared to other climate zones. **Polar deserts**, which are found in parts of Antarctica, receive very little precipitation, similar to hot deserts but in freezing conditions.

In addition to these major climate zones, there are specific subzones, such as **deserts** and **highland climates**. Deserts, found in both tropical and temperate regions, are defined by **low precipitation** rather than temperature. The Sahara Desert, for example, is hot, while the Gobi Desert experiences cold winters. **Highland climates** occur in mountainous regions where altitude affects temperature. In highlands, temperatures drop as altitude increases, creating cooler climates at the top of mountains compared to the base, even within the same latitude.

Atmospheric Circulation and Weather Patterns

Atmospheric circulation refers to the large-scale movement of air around Earth that redistributes heat and moisture, shaping weather patterns and climate zones. The primary driver of atmospheric circulation is the uneven heating of Earth's surface by the Sun. Because the equator receives more direct sunlight than the poles, temperature differences arise, causing air to move in patterns that help regulate temperature across the globe.

One of the fundamental components of atmospheric circulation is the **Hadley cell**, which dominates the tropical regions. Warm air rises at the equator due to intense solar heating. As the air rises, it cools and water vapor condenses, forming clouds and generating heavy rainfall. This is why tropical regions, such as the Amazon and Congo basins, experience high levels of precipitation. After rising and releasing moisture, the now drier and cooler air moves poleward. Around 30° latitude in both hemispheres, this air sinks, creating areas of high pressure. The descending air is dry, leading to the formation of many of the world's deserts, such as the Sahara and the Arabian Desert.

The **Ferrel cell** operates between 30° and 60° latitude. Air within this cell moves from areas of high pressure at 30° latitude toward low-pressure regions near 60°. As it travels, this air interacts with the **westerlies**, winds that blow from west to east, which are a defining feature of mid-latitude weather systems. The westerlies are responsible for moving weather systems across regions like North America and Europe, creating the frequent changes in weather observed in temperate climates. Storms, frontal systems, and cyclones are common in these regions, driven by the movement of air masses with differing temperatures and moisture levels.

In the polar regions, the **polar cell** governs atmospheric circulation. Cold air sinks at the poles, creating high-pressure zones. This cold air flows toward lower latitudes, where it meets warmer air from the Ferrel cells around 60° latitude. The interaction between these warm and cold air masses forms the **polar front**, a zone of low pressure where storms and cyclones frequently develop. This process is essential for maintaining the climate balance between the polar and temperate regions.

A key factor influencing the movement of air in the atmosphere is the **Coriolis effect**. Due to Earth's rotation, air doesn't move in straight lines but is deflected. In the Northern Hemisphere, air is deflected to the right, while in the Southern Hemisphere, it is deflected to the left. This deflection creates the characteristic wind patterns we observe, such as the **trade winds**, which blow from east to west in the tropics, and the westerlies, which dominate mid-latitudes. The Coriolis effect also determines the rotation of large-scale weather systems. In the Northern Hemisphere, cyclones rotate counterclockwise, while in the Southern Hemisphere, they rotate clockwise.

The interaction between atmospheric circulation and weather patterns is especially visible in **monsoon systems**. Monsoons are seasonal wind patterns that result in heavy rainfall during certain parts of the year. In places like South Asia, monsoons occur when the land heats up faster than the ocean during summer. Warm air over the land rises, drawing in moisture-laden air from the ocean. This moisture-laden air then rises, cools, and releases intense rainfall, leading to the monsoon season. In winter, the process reverses, leading to drier conditions.

Another important aspect of atmospheric circulation is the **jet stream**, a fast-moving band of air located near the boundary between the Ferrel and polar cells. The jet stream has a critical role in steering weather systems across the globe. It influences the development and movement of low-pressure systems, storm fronts, and even extreme weather events like heatwaves and cold snaps. Changes in the position or strength of the jet stream can have significant effects on regional weather patterns. For instance, a weakening of the jet stream can allow cold Arctic air to move southward, leading to colder-than-normal conditions in temperate regions.

Tropical cyclones are another weather phenomenon shaped by atmospheric circulation. These powerful storms form in tropical regions where warm ocean water fuels the development of intense low-pressure systems. As air rises rapidly in the center of the storm, it draws in more air from the surrounding area, creating the characteristic rotation of the cyclone. The Coriolis effect causes this rotation, which intensifies as the storm grows. Cyclones, hurricanes, and typhoons are all examples of tropical cyclones, depending on their location.

The Role of Oceans in Climate Regulation

Oceans cover about 71% of Earth's surface and are important in regulating the planet's climate. They act as vast heat sinks, absorbing and redistributing solar energy, and influence weather patterns, atmospheric circulation, and the global climate system. One of the key ways oceans regulate climate is through their ability to store and transport heat. Water has a high heat capacity, meaning it can absorb a significant amount of heat without a large increase in temperature. This allows oceans to absorb excess heat from the atmosphere, particularly from greenhouse gas emissions, and store it. Over 90% of the excess heat from global warming has been absorbed by the oceans.

Ocean currents are important in distributing this heat around the globe. Currents like the **Gulf Stream** and the **North Atlantic Drift** transport warm water from the equator toward the poles, while colder water moves from the poles back to the equator. This movement helps moderate global temperatures by transferring heat across different regions. For example, the Gulf Stream keeps northern Europe warmer than it would otherwise be given its latitude. Without this redistribution of heat, temperature differences between the equator and the poles would be much more extreme, and weather patterns would be more volatile.

The oceans also regulate climate by acting as **carbon sinks**. Phytoplankton, tiny photosynthetic organisms in the upper layers of the ocean, absorb carbon dioxide (CO_2) from the atmosphere during photosynthesis. This process not only removes CO_2 from the atmosphere, helping to reduce the greenhouse effect, but also produces oxygen. A significant portion of the carbon absorbed by the ocean eventually sinks to deeper layers, where it can be stored for hundreds or even thousands of years. This **biological pump** effectively transfers carbon from the atmosphere to the deep ocean, helping to moderate atmospheric CO_2 levels.

Oceans also absorb CO_2 directly from the atmosphere. When carbon dioxide dissolves in seawater, it forms carbonic acid, which lowers the pH of the water. This process, known as **ocean acidification**, is a growing concern because it makes it harder for marine organisms like corals and shellfish to build their calcium carbonate shells. While this direct absorption of CO_2 helps mitigate global warming, it comes at the cost of damaging marine ecosystems, which are vital for maintaining biodiversity and supporting fisheries.

The **El Niño-Southern Oscillation (ENSO)** is a key ocean-atmosphere interaction that has a profound impact on global climate. During an **El Niño** event, the trade winds in the Pacific weaken, and warm water that is usually pushed toward Southeast Asia shifts eastward, warming the central and eastern Pacific. This disrupts normal weather patterns, leading to increased rainfall in some areas, like the western coast of the Americas, and droughts in others, such as Southeast Asia and Australia. **La Niña**, the opposite phase of ENSO, occurs when cooler-than-normal ocean temperatures develop in the central Pacific, intensifying normal weather patterns. These oscillations demonstrate how changes in ocean temperatures can influence global weather and climate conditions.

In polar regions, the interaction between the oceans and sea ice has a significant influence on the climate. The **polar oceans** help regulate Earth's albedo, or reflectivity, by controlling the extent of sea ice. Ice reflects much of the sunlight that hits it, keeping polar regions cool. As sea ice melts due to global warming, darker ocean water is exposed, which absorbs more solar energy and accelerates the warming process. This feedback loop contributes to more rapid warming in the Arctic, a phenomenon known as **Arctic amplification**.

Oceanic heat content also affects the development of **hurricanes and typhoons**. These storms draw energy from warm surface waters, which fuel their intensity. As ocean temperatures rise due to global warming, hurricanes are becoming more powerful and destructive, with higher wind speeds and more intense rainfall.

Oceans also influence the **hydrological cycle**, as they are the primary source of water vapor for the atmosphere. Evaporation from the ocean's surface supplies moisture to the atmosphere, which then falls as precipitation. Changes in ocean temperature can alter evaporation rates, impacting rainfall patterns and the distribution of freshwater resources globally.

Climate Feedback Mechanisms: Positive and Negative

Climate feedback mechanisms are processes that can either amplify or dampen the effects of climate change. **Positive feedback mechanisms** amplify the initial change, leading to a more significant impact, while **negative feedback mechanisms** counteract the change, helping to stabilize the climate system. These feedbacks are central to understanding the complexity of climate systems and predicting how they will respond to increasing greenhouse gas concentrations.

One of the most well-known **positive feedback mechanisms** is the **ice-albedo feedback**. Albedo refers to the reflectivity of Earth's surface. Ice and snow have a high albedo, meaning they reflect a large portion of incoming solar radiation back into space, keeping the surface cool. As global temperatures rise and ice melts, the exposed darker surfaces, such as ocean water or land, have a much lower albedo and absorb more solar energy. This leads to further warming, which causes more ice to melt, continuing the cycle. This feedback is particularly pronounced in the Arctic, where the loss of sea ice is accelerating due to global warming.

Another significant positive feedback mechanism is the **water vapor feedback**. As the atmosphere warms, it can hold more water vapor, which is itself a potent greenhouse gas. Increased water vapor traps more heat in the atmosphere, further raising temperatures. This additional warming causes even more water vapor to evaporate, creating a cycle of increased warming. Since water vapor is the most abundant greenhouse gas, this feedback greatly amplifies the warming caused by CO_2 and other gases.

In the oceans, **methane hydrates**, which are compounds of methane trapped in ice-like structures under the seafloor, present another potential positive feedback. As ocean temperatures rise, these hydrates can destabilize, releasing methane—a potent greenhouse gas—into the atmosphere. This release of methane could lead to rapid warming, creating a feedback loop that exacerbates climate change.

On the other hand, **negative feedback mechanisms** work to moderate or reduce the effects of warming. One key negative feedback is the **radiative feedback**. As Earth's surface warms, it emits more infrared radiation (heat) back into space. This radiation loss helps to balance the energy gained from the Sun, preventing runaway warming. The efficiency of this process depends on factors such as the concentration of greenhouse gases in the atmosphere, which can limit how much heat escapes.

Cloud feedback can be both positive and negative, depending on the type and altitude of the clouds. Low clouds tend to reflect more sunlight, cooling the surface and acting as a negative feedback. High-altitude clouds, like cirrus clouds, trap heat and contribute to warming, acting as a positive feedback. The net effect of cloud feedback is complex and remains one of the largest uncertainties in climate models, as it depends on changes in cloud cover, cloud thickness, and cloud type.

Another important negative feedback is the **carbon cycle feedback** involving plants and oceans. Plants absorb CO_2 during photosynthesis, helping to reduce the amount of greenhouse gases in the atmosphere. As CO_2 levels rise, some plant species may grow more rapidly, absorbing more carbon in the process and acting as a carbon sink. Similarly, the oceans absorb a large amount of CO_2, helping to buffer the effects of increased emissions. However, there are limits to how much CO_2 plants and oceans can absorb, and changes in temperature can affect the efficiency of these natural carbon sinks.

Ocean circulation feedback is another negative feedback mechanism. In particular, the **thermohaline circulation**, also known as the ocean conveyor belt, helps to regulate global temperatures by moving warm and cold water around the planet. As polar ice melts, the influx of freshwater could potentially disrupt this circulation, slowing down the transport of warm water to higher latitudes. This would reduce heat transfer and could lead to localized cooling in regions like northern Europe, despite overall global warming.

In ecosystems, **vegetation feedbacks** can also have a role in regulating the climate. Increased plant growth due to higher levels of CO_2 can increase the amount of carbon sequestered in forests and soils. However, deforestation and land-use changes can disrupt this balance, turning forests from carbon sinks into carbon sources, which would act as a positive feedback mechanism.

Understanding the mix of these positive and negative feedback mechanisms is critical for predicting future climate scenarios. Positive feedbacks can accelerate climate change, while negative feedbacks provide a moderating influence, helping to stabilize the climate system. However, as the climate system becomes more strained by human activities, the balance between these feedbacks could shift, potentially leading to more extreme and unpredictable outcomes.

CHAPTER 5: BIODIVERSITY AND CONSERVATION

Levels of Biodiversity: Genetic, Species, Ecosystem

Biodiversity refers to the variety of life on Earth and can be understood at three distinct levels: **genetic diversity**, **species diversity**, and **ecosystem diversity**. Each of these levels contributes to the overall health, stability, and resilience of natural environments, making them essential for ecosystems to function properly. The balance and interaction between these levels of biodiversity support life on the planet and help maintain the complex systems that regulate climate, air, and water.

Genetic diversity refers to the variation of genes within a species. It's the range of genetic material present in the DNA of individuals, populations, and species. This diversity is critical because it enables populations to adapt to changing environmental conditions, diseases, and other pressures. Without genetic diversity, species are less able to survive when faced with challenges, such as climate change or habitat destruction.

For example, imagine a population of plants that experiences a drought. If there is enough genetic diversity in this population, some individuals might have traits—such as deeper roots or more efficient water use—that allow them to survive the drought. These plants would pass on their genes to the next generation, helping the population continue to thrive in drier conditions. On the other hand, if the population had low genetic diversity, there might not be enough variation in drought tolerance, leading to mass die-offs. This adaptability provided by genetic diversity is what allows species to evolve over time.

In humans, genetic diversity is what accounts for differences in characteristics such as skin color, blood types, and disease resistance. A population with more genetic diversity is generally healthier and better able to resist widespread disease outbreaks. In agriculture, maintaining genetic diversity in crops is also important for food security. Modern farming practices, which often rely on a small number of genetically uniform crops, are vulnerable to pests or diseases that could wipe out entire fields. Diverse crops, however, can help prevent total loss by having a wider range of resistance to these threats.

Species diversity refers to the variety of species within a given area or ecosystem. It's what most people think of when they hear the term biodiversity. The presence of a wide range of species creates a more complex and stable ecosystem, as each species has a specific role, or **niche**, in maintaining the system's balance. Species diversity is important because it enhances the resilience of ecosystems to disturbances like disease outbreaks, climate shifts, or human-induced changes.

In an ecosystem, different species often interact in ways that help sustain the environment. For instance, **keystone species**, such as sea otters or wolves, have a disproportionately large impact on their ecosystems. When sea otters are present, they keep sea urchin populations in check, which in turn allows kelp forests to thrive. Without otters, sea urchins would overgraze the kelp, leading to the collapse of the entire ecosystem. Similarly, predators like wolves control prey populations, which affects plant growth and, ultimately, the structure of the entire landscape. Losing a single species can lead to cascading effects throughout the ecosystem, disrupting food webs and other ecological processes.

Ecosystem diversity refers to the variety of ecosystems found within a region or across the planet. An ecosystem is a community of living organisms interacting with their physical environment, and ecosystems can range from deserts to forests to coral reefs. Each ecosystem offers different conditions, such as climate, soil types, and available resources, which support distinct groups of species and biological processes.

Ecosystem diversity is vital because it supports the wide range of services that ecosystems provide. These services include air and water purification, carbon sequestration, nutrient cycling, and the regulation of weather patterns. For example, wetlands act as natural water filters, trapping pollutants and excess nutrients before they reach rivers and lakes. Forests absorb carbon dioxide, helping to regulate the global climate. Coastal ecosystems, like mangroves and coral reefs, protect shorelines from erosion and storm surges. When ecosystem diversity declines, the loss of these services can have far-reaching impacts on human well-being.

In addition to the direct benefits ecosystems provide, they also support cultural and recreational values. Biodiverse ecosystems attract ecotourism, which can be a significant economic driver for many regions. Moreover, ecosystems hold spiritual and cultural importance for indigenous and local communities, whose identities and livelihoods are closely tied to the natural world.

The interaction between genetic, species, and ecosystem diversity is what makes ecosystems resilient and adaptable. When all three levels of biodiversity are intact, ecosystems can better withstand environmental changes and recover from disturbances. However, human activities such as deforestation, pollution, overfishing, and climate change are reducing biodiversity at all levels. Protecting and conserving biodiversity, therefore, is critical to ensuring the continued health of ecosystems and the services they provide to humanity.

Maintaining biodiversity is not just about protecting individual species—it's about preserving the complex web of life that underpins our planet's climate systems and supports all forms of life on Earth.

Threats to Biodiversity: Habitat Loss, Overexploitation

Biodiversity is under immense pressure due to human activities. Two of the most significant threats are **habitat loss** and **overexploitation**. These factors contribute to species extinction, reduce ecosystem resilience, and disrupt the balance of natural processes. Both issues are driven by economic development, population growth, and unsustainable practices, and their impacts are felt across every corner of the planet.

Habitat loss occurs when natural environments, such as forests, wetlands, and grasslands, are converted into agricultural land, urban areas, or industrial zones. Deforestation, in particular, is a major driver of habitat loss. When forests are cleared to make way for crops, livestock, or infrastructure, the species that depend on those forests are displaced or killed. The Amazon rainforest, for example, has lost millions of hectares to logging and agriculture, threatening countless species, from jaguars to rare birds and insects. Forest ecosystems are incredibly complex, supporting a high level of biodiversity. When large areas are cleared, these systems can collapse, leading to the loss of both known and yet-undiscovered species.

Urbanization is another significant factor contributing to habitat loss. As cities expand to accommodate growing populations, natural habitats are replaced with concrete, roads, and buildings. This transformation isolates wildlife populations, making it harder for them to find food, reproduce, or migrate. Fragmentation, the process where large habitats are divided into smaller, isolated patches, can be just as harmful as complete habitat destruction. When habitats are fragmented, animals and plants lose access to the resources they need to survive, and their populations can dwindle over time. The loss of connectivity between habitats also disrupts migration patterns, which are critical for many species, including birds and large mammals like elephants.

Agriculture is a major contributor to habitat destruction, especially when it involves monoculture farming or the use of chemical-intensive practices. Monoculture, where a single crop is grown over vast areas, reduces the diversity of plants and animals that can live in those areas. It also increases the need for chemical inputs like pesticides and fertilizers, which can further harm biodiversity by contaminating soil and water systems. Wetlands, which are crucial for filtering water and supporting aquatic biodiversity, are often drained for agriculture, reducing their ability to support life and maintain water quality.

Overexploitation refers to the unsustainable use of natural resources, where species are harvested faster than they can replenish. This threat affects terrestrial, marine, and freshwater ecosystems. One of the clearest examples is **overfishing**, which has led to the collapse of many fish populations. Species like bluefin tuna and Atlantic cod have been driven to critically low levels due to decades of unsustainable fishing practices. When top predators like tuna are removed from marine ecosystems, the entire food web is affected. Other species that rely on them for food, or are controlled by their predation, may either decline or proliferate in ways that destabilize the ecosystem.

Similarly, **hunting** and **poaching** pose a significant threat to terrestrial biodiversity. Iconic species like elephants, rhinos, and tigers are targeted for their tusks, horns, and skins, often driven by illegal markets. The loss of these keystone species can have ripple effects throughout ecosystems. Elephants, for example, shape their environment by uprooting trees and creating clearings, which helps maintain the balance of savanna ecosystems. Without them, the landscape can become overgrown, altering the structure of the habitat and the species it can support.

Timber harvesting is another form of overexploitation that contributes to habitat degradation. While trees are a renewable resource, unsustainable logging practices, such as clear-cutting, can devastate forest ecosystems. Clear-cutting removes entire sections of forest, leaving behind barren landscapes that take decades or even centuries to recover. During this time, the biodiversity that once thrived in these forests may be lost permanently.

Wildlife trade also fuels overexploitation, as species are captured or killed for the pet industry, traditional medicine, or ornamental purposes. The global trade in wildlife has pushed many species to the brink of extinction. Amphibians, birds, reptiles, and mammals are all affected, often with little regard for the long-term consequences on populations or ecosystems. The trade in exotic pets, for example, has severely impacted populations of tropical birds and reptiles. Removing animals from the wild not only reduces their populations but also disrupts the ecological roles they play, such as pollination, seed dispersal, or controlling insect populations.

Both habitat loss and overexploitation are compounded by other factors like **climate change** and **pollution**, creating a complex web of threats to biodiversity. Habitat destruction reduces species' ability to migrate or adapt to changing temperatures, while overexploitation weakens ecosystems, making them more vulnerable to external pressures. Once biodiversity declines beyond a certain threshold, ecosystems can lose their ability to recover, leading to long-term environmental degradation.

Efforts to combat habitat loss and overexploitation include creating **protected areas**, enforcing wildlife laws, and promoting **sustainable resource management**. However, global cooperation and shifts in how humans interact with the environment are necessary to prevent further biodiversity loss and to preserve ecosystems that provide essential services like carbon storage, clean air, and water filtration.

Conservation Strategies: In-Situ and Ex-Situ

Conservation efforts aim to protect species, ecosystems, and genetic diversity, and these efforts can be broadly divided into two main strategies: **in-situ** and **ex-situ** conservation. Both approaches are essential for maintaining biodiversity and mitigating the negative effects of habitat destruction, overexploitation, and climate change.

In-situ conservation refers to the protection of species within their natural habitats. This approach focuses on maintaining ecosystems in their original settings, allowing species to continue interacting with their environment. Examples include the establishment of **protected areas** such as national parks, wildlife reserves, and marine sanctuaries. These areas are designated to prevent human activities like logging, hunting, and mining that could threaten species and habitats.

One of the main advantages of in-situ conservation is that it preserves not just individual species but also the ecological processes and interactions that support them. Species continue to evolve naturally in their environments, maintaining the genetic diversity necessary for adaptation to environmental changes. Additionally, in-situ conservation promotes the protection of entire ecosystems, ensuring that all components, from plants to top predators, are safeguarded.

For example, the creation of national parks like Yellowstone in the U.S. has protected iconic species such as the gray wolf, bison, and grizzly bear while preserving the broader ecosystem. In tropical regions, in-situ strategies protect rainforests that are home to high levels of biodiversity, including species yet to be discovered.

Ex-situ conservation, on the other hand, involves protecting species outside of their natural habitats. This method is often used when species are critically endangered and at immediate risk of extinction. **Zoos**, **botanical gardens**, and **seed banks** are examples of ex-situ conservation strategies. In these controlled environments, species are bred, propagated, or stored to ensure their survival.

While ex-situ conservation can't fully replicate natural ecosystems, it is vital for saving species when their habitats have been degraded or destroyed. For example, **captive breeding programs** have been instrumental in reviving populations of endangered species like the California condor and the black-footed ferret. Once populations are stable, individuals can be reintroduced into their natural environments when conditions improve.

Seed banks, such as the **Svalbard Global Seed Vault**, store seeds from a wide variety of plant species, safeguarding genetic diversity and providing a backup in case of crop failure or environmental disaster. Similarly, botanical gardens have a role in ex-situ conservation by preserving plant species that are threatened in the wild.

Both in-situ and ex-situ conservation have limitations. In-situ efforts are often challenged by habitat degradation and human-wildlife conflicts, while ex-situ approaches may not fully capture the complexity of an ecosystem. However, a combination of these strategies can offer a more comprehensive approach to biodiversity conservation, ensuring that species are protected both in their natural habitats and in secure, controlled environments.

Ecosystem Services and Their Importance to Humans

Ecosystem services are the benefits that humans derive from natural ecosystems. These services are fundamental to human survival and well-being, and they are typically divided into four main categories: **provisioning services**, **regulating services**, **cultural services**, and **supporting services**.

Provisioning services refer to the direct products that ecosystems provide, such as food, water, timber, and medicinal resources. Forests, for example, provide timber for construction and fuel, while oceans and rivers supply fish and other seafood, which are essential for food security. Agricultural systems depend on healthy ecosystems to produce crops, while natural resources like fresh water, collected from rivers, lakes, and aquifers, are vital for drinking, irrigation, and sanitation. Medicinal plants found in rainforests have been the source of many life-saving drugs, such as those used to treat cancer or infections.

Regulating services are the natural processes that help regulate environmental conditions. Forests, wetlands, and oceans are important in **climate regulation** by absorbing carbon dioxide from the atmosphere, helping to mitigate the effects of climate change. Wetlands and mangroves act as natural buffers, protecting coastal areas from flooding and erosion during storms. Pollinators, like bees and butterflies, are critical for the reproduction of many crops, making them essential for agriculture. Forests also regulate water cycles by maintaining soil structure and preventing erosion, reducing the risk of landslides and preserving water quality.

Cultural services include the non-material benefits that people gain from ecosystems, such as spiritual, recreational, and aesthetic experiences. Natural landscapes, such as mountains, rivers, and forests, provide spaces for recreation and tourism, which contribute to mental health and well-being. National parks and protected areas attract millions of tourists each year, generating economic revenue and promoting the cultural value of nature. Many indigenous communities have deep spiritual connections to their natural surroundings, and these cultural ties are integral to their identities and ways of life.

Supporting services are the underlying processes that sustain ecosystems, making all other services possible. These include nutrient cycling, soil formation, and the photosynthesis that produces oxygen. For instance, decomposers like fungi and bacteria break down dead organic matter, returning nutrients to the soil, which supports plant growth. The **water cycle**, driven by evaporation and precipitation, distributes fresh water across the globe, replenishing rivers, lakes, and groundwater supplies. Biodiversity is critical to these supporting services, as the interaction between different species ensures the resilience and functioning of ecosystems.

The loss of ecosystems and the services they provide can have profound consequences for human societies. **Deforestation** can disrupt water cycles, leading to reduced rainfall and increased droughts in agricultural regions. The decline of pollinators threatens food production, potentially reducing crop yields and increasing food insecurity. **Coral reef degradation** due to climate change and pollution diminishes fish populations, affecting the livelihoods of millions of people who depend on fisheries for income and nutrition.

Moreover, the overexploitation of ecosystems, such as through overfishing or unsustainable logging, reduces the ability of these systems to provide services in the long term. As ecosystems become degraded, their capacity to regulate climate, purify water, and protect against natural disasters is diminished, leaving human populations more vulnerable to environmental and economic instability.

Community-Based Conservation Approaches

Community-based conservation (CBC) approaches involve local communities in the management and protection of natural resources. This strategy recognizes that people who live closest to natural resources have a direct stake in their conservation, and they possess valuable knowledge about the local environment. In many cases, engaging local communities leads to more sustainable and effective conservation outcomes.

One key element of community-based conservation is the **decentralization of management**. Rather than relying solely on government agencies or external organizations, CBC places decision-making power in the hands of local communities. This can lead to better management of forests, wildlife, and fisheries because the people who rely on these resources for their livelihoods are more motivated to protect them. For example, in parts of East Africa, local communities have been given the authority to manage wildlife through **community conservancies**, which allow them to benefit from tourism revenue while conserving important habitats for species like elephants and lions.

Traditional ecological knowledge (TEK) is another critical component of CBC. Indigenous and local communities often have a deep understanding of the ecosystems in which they live, based on generations of observation and interaction with the land. This knowledge can be used to develop sustainable resource management practices that are better suited to the local context than externally imposed methods. For instance, many indigenous groups practice rotational farming, which allows forests to regenerate and maintains soil fertility, reducing the need for destructive agricultural expansion.

CBC also emphasizes the **economic benefits** of conservation. When communities see direct economic benefits from protecting natural resources, they are more likely to support conservation efforts. Ecotourism, for example, can provide jobs and income for local people while promoting the protection of wildlife and ecosystems. In Costa Rica, community-led ecotourism initiatives have been successful in preserving rainforests and marine areas while generating income for local residents through activities like guided tours and wildlife viewing.

However, community-based conservation also faces challenges. **Conflicts over land rights**, resource use, and the distribution of benefits can arise, particularly in areas where there is competition for natural resources. It is important for CBC projects to ensure that benefits are equitably distributed and that local communities have a genuine say in decision-making processes. Additionally, there needs to be a balance between conservation goals and the development needs of local populations to ensure long-term sustainability.

CHAPTER 6: POPULATION ECOLOGY

Population Dynamics: Growth, Density, Distribution

Population dynamics refers to the changes in population size, density, and distribution over time. These changes are influenced by factors like birth rates, death rates, immigration, and emigration. Understanding how populations grow, how densely they are packed in a given area, and how they are distributed helps ecologists predict population trends and their impacts on ecosystems.

Population growth is one of the key aspects of population dynamics. Populations can grow in different ways depending on the availability of resources, environmental conditions, and species-specific traits. Under ideal conditions, populations can experience **exponential growth**, where the population size increases rapidly because every individual has abundant resources, like food and space. In this scenario, the growth rate of the population accelerates as more individuals are born and reproduce. For example, bacteria in a nutrient-rich environment can double their numbers every few hours because they are not limited by resources, allowing their population to expand rapidly.

However, exponential growth doesn't occur indefinitely. As a population grows, resources like food, water, and space become limited. When this happens, population growth slows and eventually reaches a **carrying capacity**—the maximum number of individuals that an environment can sustain over time. This results in **logistic growth**, where the population size levels off as it approaches the carrying capacity. For instance, a population of deer in a forest will grow quickly if there are plenty of food sources, but as the number of deer increases, food becomes scarcer, leading to a slowdown in population growth until the population stabilizes.

Factors like competition for resources, predation, and disease often regulate population growth. When populations exceed the carrying capacity, they may experience **population crashes** due to the overuse of resources, leading to higher mortality rates. Predators, for example, may reduce the population of their prey if the prey population grows too large. On the other hand, favorable environmental conditions, such as a mild climate or an abundance of food, can lead to population booms.

Population density refers to the number of individuals in a given area or volume. High population density means many individuals are packed into a relatively small area, while low density indicates that individuals are more spread out. Population density affects how individuals interact with one another and with their environment. In high-density populations, competition for resources like food, space, and mates can be intense. This can lead to stress and increased vulnerability to disease, both of which can reduce population growth. For example, a large number of animals living in close quarters may experience higher rates of disease transmission, as pathogens can spread more easily in dense populations.

Low-density populations, in contrast, may face challenges in finding mates, which can limit reproduction. Species that are solitary or territorial often have lower population densities because individuals require larger areas to meet their needs. For instance, animals like tigers or wolves maintain large territories, and as a result, their population densities are relatively low compared to species that live in groups, like birds in a flock or fish in a school.

Population distribution describes how individuals in a population are spaced across their environment. There are three main types of distribution patterns: **clumped**, **uniform**, and **random**. Each of these patterns reflects how individuals use the available resources and interact with one another.

In **clumped distribution**, individuals are grouped together in patches. This is the most common type of distribution in nature and often occurs where resources like food or water are unevenly distributed. For example, animals like elephants tend to cluster around watering holes in dry environments. Clumped distribution also occurs when species benefit from group living, such as for protection from predators. Herds of animals, like buffalo or zebras, group together for safety in numbers, reducing the likelihood of being singled out by predators.

Uniform distribution is less common and occurs when individuals are evenly spaced. This pattern often results from competition for resources or territorial behavior. For example, some bird species, such as penguins, exhibit uniform distribution because they need specific distances between nests to avoid aggression and ensure that each bird has access to adequate space and resources. Plants that release chemicals to inhibit the growth of other plants around them, a phenomenon known as allelopathy, may also show uniform distribution as they maintain space between individuals.

Random distribution occurs when the position of one individual is independent of others. This pattern is rare in nature because most environmental factors, like resource availability or social behavior, tend to influence the spacing of individuals. However, it can occur in species where resources are plentiful and evenly distributed, and there is little competition or social interaction among individuals. For instance, some plants that rely on wind dispersal of seeds might show a random distribution, as the seeds land wherever the wind carries them.

Overall, population dynamics—encompassing growth, density, and distribution—are shaped by the interactions between organisms and their environment. Whether a population is growing, shrinking, or stable depends on the balance of factors like birth rates, death rates, and resource availability. By studying these dynamics, ecologists gain insights into how populations adapt to environmental changes and how they impact the ecosystems they inhabit.

Carrying Capacity and Limiting Factors

Carrying capacity refers to the maximum number of individuals of a species that an environment can support indefinitely without degrading the ecosystem. It's a concept central to population ecology, helping explain why populations don't grow indefinitely and how they stabilize at a certain level. The carrying capacity of an environment is determined by several **limiting factors**—environmental conditions that prevent populations from growing beyond a certain point. These factors can be classified as **density-dependent** or **density-independent**, and they are important in regulating populations by influencing birth rates, death rates, and migration.

One of the most important aspects of carrying capacity is its dynamic nature. The carrying capacity for a population can change over time due to variations in resource availability, environmental conditions, and species interactions. For example, a region's carrying capacity for deer may increase during wet seasons when food is abundant but decrease during droughts when resources become scarce. In essence, carrying capacity is not a fixed number but a balance between the needs of the population and the capacity of the environment to provide for those needs.

Density-dependent limiting factors are those that increase in intensity as a population grows and becomes more crowded. These factors include competition for resources like food, water, and shelter, as well as predation, disease, and parasitism. In dense populations, individuals compete more fiercely for limited resources, which can lead to higher mortality rates and lower reproduction rates. For example, in a forest where food is scarce, deer may struggle to find enough to eat, leading to malnutrition, weaker offspring, and higher death rates, which ultimately slow population growth.

Predation is another density-dependent factor that can regulate population size. As prey populations increase, predators may find it easier to locate and capture prey, which can lead to a rise in predator populations. As predators consume more prey, the prey population declines, and this can create a feedback loop that helps keep the population in check. For example, in a predator-prey relationship between wolves and elk, a growing elk population may lead to an increase in wolf numbers. As wolves hunt more elk, the elk population declines, which in turn reduces the food available for wolves, eventually stabilizing both populations.

Disease and parasitism also become more prevalent in denser populations. Pathogens and parasites spread more easily when individuals are packed closely together, leading to outbreaks that can reduce population size. For example, wildlife populations like birds or fish can experience die-offs due to diseases that spread rapidly through dense colonies. This is nature's way of regulating population size to avoid overuse of resources.

In contrast, **density-independent limiting factors** affect populations regardless of their density. These factors are often abiotic, such as weather events, natural disasters, or human activities like pollution and habitat destruction. For instance, a severe drought can reduce the availability of water and food for all individuals in an ecosystem, regardless of how many animals live there. Similarly, a wildfire may decimate a forest population by destroying habitat and food sources, regardless of how dense the population was before the fire.

Changes in climate or extreme weather events, such as hurricanes or floods, can drastically lower an environment's carrying capacity by altering the availability of resources or physically displacing populations. For example, a flood may reduce the amount of land available for grazing animals like cattle or wild herbivores, thereby reducing the environment's ability to support large numbers of these animals.

Human activities can also be significant **limiting factors**. Overfishing, deforestation, and pollution can all reduce the carrying capacity of natural ecosystems. For example, overfishing in marine environments can reduce fish stocks to levels where they are no longer able to sustain their populations, effectively lowering the carrying capacity of the ocean for these species. Habitat destruction, caused by urbanization or agricultural expansion, reduces the amount of available space and resources for wildlife, shrinking the carrying capacity of the affected ecosystems.

In some cases, populations can temporarily exceed their carrying capacity, a phenomenon known as **overshoot**. This occurs when the environment can no longer sustain the population at its current level, leading to resource depletion, habitat degradation, and a sharp population decline, often referred to as a **crash**. For example, if a population of deer in a forest exceeds the carrying capacity due to an abundance of food one year, they may overgraze the vegetation, leading to food shortages in subsequent years. This would cause the population to crash until it reaches a level that the degraded environment can support.

It's important to recognize that carrying capacity is not only about food or water—it includes all essential resources and environmental factors that species depend on for survival. This can range from the availability of mates and shelter to the quality of soil or water in an ecosystem. As populations approach carrying capacity, limiting factors like waste accumulation or decreased resource quality can also have a role in slowing population growth.

Human intervention can sometimes temporarily increase the carrying capacity of an environment through activities such as agriculture, irrigation, or wildlife management. However, these changes can have long-term ecological consequences, often degrading ecosystems and reducing their ability to sustain populations over time. For example, intensive farming practices can increase crop yields in the short term but lead to soil depletion and erosion, reducing the land's ability to support agriculture in the future.

Human Population Growth and Environmental Impact

Human population growth has been a dominant force in shaping the environment, particularly since the industrial revolution. The global population has surged from around 1 billion in 1800 to over 7.9 billion today, with significant implications for ecosystems, resource use, and climate. Rapid population growth places increasing pressure on the planet's natural resources and contributes to environmental degradation in various forms.

One of the most visible impacts of human population growth is the **overexploitation of natural resources**. As populations grow, the demand for food, water, energy, and raw materials increases. This has led to widespread deforestation, overfishing, and the depletion of freshwater resources. For example, forests are cleared at alarming rates to make room for agriculture, infrastructure, and urban development. This not only destroys habitats but also reduces biodiversity and contributes to climate change by removing carbon sinks that absorb CO_2 from the atmosphere.

The growth of agriculture to feed the growing human population has particularly severe environmental consequences. **Intensive farming** practices, which involve the use of chemical fertilizers, pesticides, and irrigation, can lead to soil degradation, water pollution, and the depletion of aquifers. Runoff from agricultural lands often carries excess nutrients into rivers and lakes, causing **eutrophication**, where algae blooms deplete oxygen in the water, harming aquatic life. In regions like the Amazon, large swaths of forest are cleared to grow crops like soy or to raise cattle, further driving habitat loss and carbon emissions.

Human population growth also leads to **increased energy consumption**, much of which is still derived from fossil fuels like coal, oil, and natural gas. The burning of these fuels releases significant amounts of carbon dioxide into the atmosphere, contributing to **global warming** and climate change. As more people require transportation, electricity, and heating, the demand for energy continues to rise, leading to higher greenhouse gas emissions. The result is a warming planet with more extreme weather events, rising sea levels, and shifts in global climate patterns.

Urbanization is another significant environmental impact of human population growth. As more people move into cities, the demand for infrastructure and housing grows, leading to the expansion of urban areas into natural habitats. Urban sprawl not only consumes land but also increases air and water pollution. Cities are hotspots for greenhouse gas emissions, especially from transportation and industry, and they contribute to the **urban heat island effect**, where urban areas are significantly warmer than their rural surroundings due to human activities and heat-absorbing materials like concrete and asphalt.

The **waste** generated by a growing human population also poses environmental challenges. As consumption increases, so does the production of waste, including plastics, electronic waste, and hazardous materials. Much of this waste ends up in landfills or in the natural environment, particularly in the oceans, where plastic pollution has become a critical issue. Marine ecosystems are particularly vulnerable, as animals can ingest or become entangled in plastic debris, leading to injury or death.

The sheer scale of human population growth exacerbates **climate change** by amplifying the demand for resources and energy. More people mean more carbon emissions from transportation, agriculture, and industry. The combined effects of deforestation, industrial emissions, and urbanization contribute to the buildup of greenhouse gases, intensifying global warming. Climate change, in turn, affects the availability of resources, creates harsher living conditions in some regions, and drives species extinction.

Efforts to mitigate the environmental impact of human population growth focus on promoting **sustainable development**. This involves using resources more efficiently, reducing waste, and transitioning to renewable energy sources like solar and wind power. **Family planning programs** that provide access to education and reproductive health services are also critical for stabilizing population growth, particularly in developing countries where birth rates remain high.

Population Control Strategies: Policies and Education

Population control in ecology refers to the methods used to regulate the size of a population, ensuring that it remains within sustainable limits. Overpopulation can lead to issues such as resource depletion, habitat destruction, and the decline of species. Ecological population control can occur naturally or be managed artificially by human intervention. Each strategy has its own ecological, ethical, and practical considerations.

Natural Population Control

Natural population control mechanisms regulate population size without human interference. These are often categorized as **density-dependent** and **density-independent** factors.

Density-dependent factors become more effective as population density increases. For example, as population size grows, **competition** for resources like food, water, and shelter intensifies. In environments where resources are limited, individuals may struggle to survive and reproduce, slowing population growth. **Predation** is another density-dependent factor. Predators tend to target species with high population densities because prey becomes more abundant and easier to catch. In such cases, predation helps regulate prey populations, keeping them in check. **Disease** is also more likely to spread rapidly in dense populations, leading to higher mortality rates. Pathogens spread more easily when individuals are crowded, further limiting population growth.

Density-independent factors, on the other hand, affect population size regardless of density. These factors include **natural disasters**, such as floods, fires, or droughts, which can devastate populations regardless of their size. For example, a severe wildfire can drastically reduce a population in a region by destroying its habitat, irrespective of the population's previous density. **Climate change** is another density-independent factor, as shifts in temperature and precipitation patterns can alter the availability of resources, potentially leading to declines in population numbers.

Artificial Population Control

Humans often intervene to manage populations for wildlife conservation, pest control, or even ecosystem restoration. Artificial population control methods can include **culling, contraception, habitat manipulation,** and **translocation**.

Culling involves the selective killing of individuals within a population to reduce its size. This method can be controversial, particularly in terms of ethical considerations, but it may be necessary to prevent overgrazing or control disease outbreaks. Culling is often used in wildlife management where a species' overpopulation threatens ecosystem stability or biodiversity.

Contraception is a more humane alternative to culling, preventing reproduction without directly reducing population size through death. While contraception is seen as a gentler method, it can be expensive and difficult to implement on a large scale, especially in wild populations. This method is more commonly used in confined settings, such as national parks or zoos, where controlling the population is essential.

Habitat manipulation involves altering the environment to make it less suitable for the target species. This might include removing food sources, destroying nesting sites, or even introducing predators. For example, altering the landscape to reduce the availability of a key resource may cause a population to decline naturally. However, this method can have unintended consequences on other species and requires careful planning.

Translocation is the process of moving individuals from one location to another. This strategy is often used when populations are overcrowded in one area, or to establish new populations in areas where a species may have become extinct. Translocation can be successful in creating viable populations in new areas but can also fail if the introduced individuals struggle to adapt to the new environment or if it disrupts the existing ecosystem.

Choosing the Right Strategy

The choice of population control strategy depends on several factors, including the species being managed, the reason for control, and the potential impacts on the broader ecosystem. **Population dynamics**, or the factors influencing population growth and decline, must be carefully studied to avoid unintended consequences. A balance between controlling population sizes and maintaining ecosystem health is essential. Ethical considerations, such as the welfare of the animals involved and public perception, also have a significant role in determining which strategies are most appropriate and acceptable.

CHAPTER 7: SOIL AND LAND USE

Soil Formation and Composition

Soil forms through a complex interaction of physical, chemical, and biological processes that break down rocks and organic matter over time. It is a dynamic, living medium that supports plant life, and its composition reflects the environment where it forms. The process of **soil formation**, or **pedogenesis**, begins with the weathering of rocks and the accumulation of organic matter from dead plants and animals. This mixture of inorganic minerals and organic material, along with water and air, creates the rich medium that sustains ecosystems and agriculture.

Weathering is the first step in soil formation, and it occurs in two main ways: **physical weathering** and **chemical weathering**. Physical weathering breaks rocks into smaller particles through mechanical processes like freezing and thawing, wind abrasion, and the expansion of plant roots. For example, when water enters cracks in rocks and freezes, it expands, causing the rock to break apart. Similarly, plant roots can grow into rock crevices, exerting pressure that splits the rock.

Chemical weathering alters the minerals in rocks through reactions with water, oxygen, and acids. Rainwater, which contains dissolved carbon dioxide, forms a weak acid (carbonic acid) that can dissolve minerals like feldspar and limestone. Over time, this process transforms hard rocks into softer materials like clay. In warm, humid climates, chemical weathering happens more quickly, contributing to faster soil formation.

As weathering breaks down rock into smaller particles, **organic matter** from plants and animals mixes in. This organic matter is essential to soil composition and forms the basis for **humus**, a dark, nutrient-rich substance that improves soil fertility. When plants shed leaves or die, microorganisms like bacteria and fungi decompose the organic material, releasing nutrients like nitrogen, phosphorus, and potassium back into the soil. Earthworms, insects, and other soil organisms also have a role in breaking down organic matter and aerating the soil, creating pathways for air and water to move.

The composition of soil is typically divided into four main components: **mineral particles**, **organic matter**, **water**, and **air**. Each component has a unique role in supporting plant growth and maintaining soil health.

1. **Mineral particles** make up the bulk of soil and come from weathered rock. These particles are classified by size into **sand**, **silt**, and **clay**. Sand particles are the largest and create a coarse texture, allowing water to drain quickly. Clay particles, on the other hand, are much smaller and have a fine

texture that holds water and nutrients more effectively. Silt falls in between, providing a balanced texture that combines the drainage properties of sand with the water-holding capacity of clay. The ratio of sand, silt, and clay in a soil determines its **texture**, which influences how well it holds water, drains, and supports root growth. **Loam**, a soil with roughly equal parts sand, silt, and clay, is often considered ideal for agriculture because it balances water retention and drainage.

2. **Organic matter** is the decomposed remains of plants, animals, and microorganisms. Although it makes up only a small percentage of the soil by volume, it is critical for soil fertility. Organic matter binds soil particles together, improving structure, and it holds nutrients that plants can absorb through their roots. Humus, a stable form of organic matter, enhances soil's ability to retain moisture and nutrients, making it more fertile.

3. **Water** is essential for plant growth and acts as a medium for transporting nutrients within the soil. Water enters the soil through precipitation and infiltration, moving through pore spaces between soil particles. Soils that are too sandy drain water too quickly, making it unavailable to plants, while clay-rich soils can become waterlogged, depriving roots of oxygen. The ideal soil retains enough water for plants while allowing excess water to drain away.

4. **Air** fills the spaces between soil particles not occupied by water. Plants need air in the soil to obtain oxygen for respiration, a critical process for root health. Soil organisms, like earthworms and microbes, also require oxygen to break down organic material and recycle nutrients. Well-structured soils have good **porosity**, meaning there are enough spaces between particles for air to circulate and water to flow.

The process of soil formation is influenced by several factors, including **climate, parent material, topography, biological activity**, and **time**. In areas with warm temperatures and abundant rainfall, soils form more quickly than in dry or cold regions. **Parent material**, the original rock from which soil forms, affects the soil's mineral content and texture. For example, granite weathers into coarse, sandy soils, while limestone produces finer, clay-rich soils. **Topography**, or the shape of the land, also has a role. Soils on steep slopes may erode more easily, while flat areas allow more water infiltration and soil accumulation. **Biological activity**, including the action of plants, animals, and microorganisms, accelerates the breakdown of organic material and the mixing of soil layers. Finally, **time** is a crucial factor—soil formation is a slow process that can take hundreds or thousands of years.

Soil Erosion and Degradation

Soil erosion and degradation are critical issues that threaten the health of ecosystems, agriculture, and water systems globally. Erosion is the process by which the top layer of soil is removed by natural forces like wind and water, while degradation refers to the decline in soil quality due to factors like chemical contamination, overuse, and loss of organic matter. Together, they reduce the soil's ability to support plant growth, disrupt natural cycles, and lead to long-term environmental damage.

Soil erosion occurs when soil is loosened and transported by wind, water, or human activity. This process is often accelerated by human activities such as deforestation, agriculture, and urban development. When vegetation is removed from the land, the protective cover that anchors the soil is lost, making it vulnerable to being washed or blown away. **Water erosion** is especially common in areas with heavy rainfall or steep slopes. As rainwater flows across the surface, it carries away the topsoil, which is rich in nutrients and organic matter. This process can lead to **rill erosion**, where small channels form in the soil, or **gully erosion**, where larger trenches develop. Over time, these eroded areas become less fertile and more difficult to cultivate.

Wind erosion is another significant issue, particularly in dry and arid regions. When the soil surface is exposed and dry, wind can pick up loose particles and transport them over long distances. This not only depletes the soil in the affected area but also creates **dust storms** that can reduce air quality and pose health risks. Large-scale dust storms, such as those seen during the Dust Bowl of the 1930s in the United States, illustrate how wind erosion can devastate agricultural land and force people to abandon their farms.

Human activities are a major driver of soil degradation. **Intensive agriculture**, especially the use of heavy machinery, monoculture cropping, and chemical fertilizers, can lead to soil compaction, loss of nutrients, and a decrease in organic matter. Over time, these practices weaken soil structure, making it more prone to erosion and less able to retain water. **Overgrazing** by livestock is another contributor, as it removes vegetation and compacts the soil, making it harder for plants to grow back.

Soil degradation also includes the loss of soil fertility due to the depletion of essential nutrients like nitrogen, phosphorus, and potassium. This often happens when soils are farmed continuously without being allowed to regenerate, or when synthetic fertilizers are used excessively. While fertilizers can temporarily boost crop yields, overuse leads to a decline in natural soil fertility and disrupts the balance of soil microorganisms that are essential for healthy soil. This degradation affects the soil's ability to support plant life, leading to lower agricultural productivity and increased reliance on chemical inputs.

Salinization is another form of soil degradation, particularly in irrigated agricultural regions. When irrigation water evaporates, it leaves behind salts that accumulate in the soil. Over time, high salt concentrations make it difficult for plants to absorb water, reducing crop yields and eventually rendering the land unusable for agriculture.

Soil compaction is caused by the repeated pressure from heavy machinery or livestock trampling. Compacted soil has fewer air spaces, making it difficult for roots to grow and for water to infiltrate the soil. This leads to increased surface runoff and erosion, as well as reduced plant growth and lower agricultural yields.

The consequences of soil erosion and degradation are severe. **Loss of topsoil** reduces the land's ability to produce crops, affecting food security for millions of people worldwide. Eroded soil often ends up in rivers and lakes, causing **siltation**, which reduces water quality and can lead to the destruction of aquatic habitats. Sedimentation can also clog waterways and reservoirs, reducing their capacity and increasing the risk of flooding.

Combatting soil erosion and degradation requires a combination of practices aimed at preserving soil structure, enhancing fertility, and protecting against erosion. **Reforestation** and planting cover crops can stabilize the soil, reducing its exposure to wind and water. **Terracing** and **contour plowing** are effective methods for slowing water runoff on slopes, preventing soil from being washed away. In agricultural areas, promoting **crop rotation** and reducing the use of chemical fertilizers can help maintain soil health. Additionally, reducing **overgrazing** through better livestock management practices can allow vegetation to recover, improving the soil's resilience.

Soil is a finite resource, and once it is lost or degraded, it can take centuries to regenerate. Preventing further soil erosion and degradation is essential to sustaining agriculture, protecting ecosystems, and ensuring the long-term health of the environment.

Agricultural Practices and Sustainable Farming

Agricultural practices have a profound impact on the environment, influencing soil health, water resources, and biodiversity. As global populations grow, the demand for food increases, putting pressure on land and ecosystems. To meet this demand sustainably, farmers and policymakers are turning to **sustainable farming practices**, which aim to maintain productivity while minimizing environmental damage and preserving resources for future generations.

Traditional agricultural practices, such as **monoculture farming**, where a single crop is grown repeatedly on the same land, have significant downsides. Monoculture depletes the soil of nutrients, making it necessary to use synthetic fertilizers to maintain crop yields. However, this leads to soil degradation over time and disrupts the balance of soil microorganisms that are vital for fertility. Monoculture also increases the vulnerability of crops to pests and diseases, as a lack of crop diversity makes it easier for pests to spread. To counteract these problems, many farmers are adopting **crop rotation**, where different crops are planted in succession to restore soil nutrients and reduce pest populations. For example, alternating between nitrogen-fixing plants like legumes and nutrient-demanding crops like corn can improve soil fertility and reduce the need for chemical fertilizers.

Another key sustainable farming practice is **conservation tillage**. In traditional farming, tilling the soil helps control weeds but also disturbs the soil structure and increases the risk of erosion. Conservation tillage minimizes soil disturbance by leaving crop residues on the surface, which protects the soil from erosion and improves water retention. This practice also helps maintain organic matter in the soil, which is critical for long-term soil health. By reducing the frequency and intensity of tilling, farmers can improve soil structure, reduce erosion, and enhance water infiltration.

Agroforestry is another sustainable practice that integrates trees and shrubs into agricultural systems. Trees help prevent soil erosion, improve water retention, and provide shade for crops, reducing the need for irrigation. Additionally, trees sequester carbon, contributing to climate change mitigation efforts. Agroforestry also increases biodiversity on farms by creating habitats for beneficial insects, birds, and other wildlife, which can help control pests naturally without relying on chemical pesticides.

Water management is a critical component of sustainable farming. **Drip irrigation** systems, for instance, deliver water directly to the plant roots, reducing water waste and minimizing evaporation. This is especially important in regions where water is scarce. In contrast to traditional flood irrigation, which can lead to waterlogging and salinization, drip irrigation uses water more efficiently, helping to maintain soil health and ensure long-term agricultural productivity.

Sustainable farming also emphasizes the reduction of chemical inputs, such as **pesticides** and **synthetic fertilizers**. Excessive pesticide use can harm non-target species, including pollinators like bees, which are essential for crop production. It can also lead to pesticide resistance in pests, making it harder to control them in the future. Instead, many farmers are turning to **integrated pest management (IPM)**, which combines biological, mechanical, and chemical methods to control pests in a more environmentally friendly way. For example, introducing natural predators like ladybugs to control aphid populations can reduce the need for chemical insecticides.

Similarly, reducing the use of synthetic fertilizers and focusing on **organic farming** methods helps maintain soil fertility without contributing to nutrient runoff, which can pollute nearby water bodies. Organic farming relies on natural fertilizers like compost and manure, which enrich the soil with organic matter and promote healthy microbial activity. By improving soil health, organic farming enhances the soil's ability to retain nutrients and water, reducing the need for additional inputs and making farms more resilient to droughts and other environmental stresses.

Another emerging approach is **precision agriculture**, which uses technology to optimize farming practices. GPS-guided machinery and soil sensors can provide detailed information about soil conditions, allowing farmers to apply water, fertilizers, and pesticides more efficiently. This reduces waste, lowers costs, and minimizes the environmental impact of farming.

Sustainable farming practices not only protect the environment but also contribute to **food security** by ensuring that agricultural land remains productive in the long term. As climate change, water scarcity, and soil degradation become increasingly urgent global challenges, sustainable farming offers a way to balance the need for food production with the need to protect ecosystems and resources for future generations.

Soil Conservation Techniques

Soil conservation techniques are vital for maintaining soil health, preventing erosion, and ensuring long-term agricultural productivity. As soil degradation becomes an increasing concern due to deforestation, intensive agriculture, and climate change, it is essential to implement strategies that protect and enhance the soil's capacity to support life. These techniques focus on preventing soil erosion, preserving soil structure, maintaining fertility, and improving water retention. Here are some of the most effective soil conservation practices used in agriculture and land management today.

Contour Plowing

Contour plowing involves plowing along the contours of a slope rather than up and down. This technique helps slow the flow of water across the land, reducing soil erosion by water runoff. By following the natural shape of the terrain, contour plowing creates barriers that trap water and allow it to infiltrate the soil instead of washing away topsoil. In areas with rolling hills or moderate slopes, this method can significantly reduce soil loss, particularly during heavy rains. Farmers can further enhance this technique by incorporating **terracing**, where steeper slopes are transformed into a series of flat, step-like areas to further reduce erosion.

Terracing

Terracing is particularly useful in mountainous or hilly regions where traditional farming would lead to severe erosion. By carving flat platforms into the hillside, terracing helps slow down water runoff and provides level areas for crops to grow. Each terrace acts as a barrier to water flow, reducing the speed at which water moves downhill, which prevents soil from being washed away. Terracing also increases the surface area for cultivation, making it a practical solution for regions with limited flat land. For example, the ancient Inca civilization used terracing extensively in the Andes mountains to support agriculture in steep terrain.

Cover Cropping

Cover cropping is the practice of planting crops specifically to protect the soil between growing seasons. These cover crops, such as clover, rye, or legumes, help prevent soil erosion by keeping the soil covered and reducing the impact of wind and rain. Additionally, cover crops enhance soil fertility by fixing nitrogen, improving organic matter, and increasing soil biodiversity. Legumes, in particular, are often used as cover crops because they have nitrogen-fixing bacteria in their root nodules, which convert atmospheric nitrogen into a form that plants can use. When these cover crops are plowed back into the soil, they act as green manure, further enriching the soil's nutrient content.

No-Till Farming

Traditional farming methods often involve tilling, or turning over the soil, to prepare it for planting. However, tilling can disrupt soil structure, increase erosion, and lead to the loss of organic matter. **No-till farming** is an alternative that minimizes soil disturbance by leaving crop residues on the surface and planting new crops directly into the undisturbed soil. This method protects the soil from erosion, improves water retention, and helps maintain soil structure. No-till farming also reduces the need for heavy machinery, cutting down on fuel use and greenhouse gas emissions. Over time, this approach can increase the organic matter in the soil, enhancing its fertility and resilience to environmental stress.

Windbreaks

Wind erosion is a significant problem in arid and semi-arid regions where high winds can strip away topsoil, leaving the land barren. **Windbreaks** are rows of trees or shrubs planted along the edges of fields to reduce the speed of the wind and protect the soil. These barriers help trap soil particles that might otherwise be blown away, preserving the topsoil and preventing desertification. Windbreaks also provide additional benefits, such as creating habitats for wildlife, reducing evaporation from soil, and improving microclimates around crops. In regions like the Great Plains of the United States, windbreaks have been crucial in preventing a recurrence of the severe dust storms that occurred during the Dust Bowl in the 1930s.

Contour Buffer Strips

Contour buffer strips are permanent strips of vegetation, typically grasses or legumes, planted along the contours of farmland. These strips help slow water runoff, reduce soil erosion, and filter out sediments and pollutants before they enter waterways. By breaking up the length of the slope, contour buffer strips act as barriers to water movement, trapping soil and organic matter that might otherwise be lost. This technique is particularly effective in areas with sloped farmland, where erosion is a significant concern. The vegetation in the buffer strips also improves soil structure, increases biodiversity, and provides habitat for beneficial insects and wildlife.

Agroforestry

Agroforestry combines trees and shrubs with crops or livestock on the same land. This practice mimics natural ecosystems, where multiple layers of vegetation work together to protect the soil. Trees in agroforestry systems help reduce soil erosion by stabilizing the soil with their roots, improving water retention, and providing shade that reduces evaporation. The leaves and organic matter from trees also enrich the soil, promoting better crop growth. Agroforestry systems can be highly diverse, incorporating fruit trees, timber species, and even medicinal plants alongside traditional crops. This diversity enhances soil health, improves resilience to climate change, and increases biodiversity on the farm.

Crop Rotation

Crop rotation involves alternating the types of crops grown on a particular piece of land from season to season. This practice helps maintain soil fertility and reduces the risk of soil degradation. Different crops have varying nutrient requirements and rooting depths, so rotating crops can prevent the depletion of specific nutrients and reduce soil erosion. For example, deep-rooted crops like alfalfa can break up compacted soil layers, improving soil structure and water infiltration, while shallow-rooted crops like wheat can use nutrients near the soil surface. Crop rotation also helps control pests and diseases, as many pests and pathogens are specific to certain crops and cannot survive if their preferred host is absent for a season.

CHAPTER 8: WATER RESOURCES AND MANAGEMENT

Global Water Distribution and Availability

Water covers about 71% of Earth's surface, but most of it is not readily available for human use. Of the total volume of water on Earth, approximately 97.5% is **saltwater** found in oceans and seas, which is not suitable for drinking, agriculture, or most industrial uses without desalination. Only about **2.5% of the Earth's water** is freshwater, and even within that small fraction, most is inaccessible to humans because it is locked away in glaciers, ice caps, or deep underground aquifers.

The available freshwater for human use is found in **surface water** bodies like rivers, lakes, and reservoirs, and in **groundwater** stored in aquifers. Surface water accounts for just **0.3%** of the world's freshwater, while groundwater comprises about **30.1%**. Groundwater is a critical resource, especially in arid regions, but it is often difficult to replenish because aquifers recharge slowly. Many regions around the world depend heavily on this resource, but **overextraction** of groundwater is leading to declining water tables in places like India, China, and the United States.

The **distribution of freshwater** is uneven across the globe. Some regions, such as the tropics and high-latitude areas, have abundant water supplies due to consistent rainfall and large river systems. For instance, the Amazon Basin in South America holds a massive portion of the world's freshwater. However, other areas, particularly arid and semi-arid regions, face significant **water scarcity**. The Middle East, North Africa, and parts of Australia are among the driest areas on Earth, relying on limited water sources such as rivers that flow through multiple countries, like the Nile or the Tigris-Euphrates river systems.

Even within countries, water availability can vary drastically. For example, in the United States, the western states like California and Nevada often experience water shortages due to prolonged droughts, while eastern states generally have more reliable rainfall and freshwater resources. **Climate change** is exacerbating these disparities by altering precipitation patterns, leading to more extreme droughts in some areas and increased flooding in others.

Another critical issue affecting global water distribution is **population growth** and **urbanization**. As populations grow, especially in developing countries, demand for water for drinking, agriculture, and industry increases. Large cities, particularly in water-scarce regions, struggle to meet the demands of their growing populations. In many cases, cities must transport water over long distances, such as Los Angeles, which relies on water from distant rivers and reservoirs.

In addition to natural water scarcity, **pollution** affects the availability of clean freshwater. Industrial waste, agricultural runoff, and untreated sewage contaminate rivers, lakes, and groundwater supplies, making them unsafe for human consumption. In countries with limited water treatment infrastructure, this pollution can exacerbate water shortages by reducing the usable water supply.

Water stress is a growing global challenge. Some regions, such as sub-Saharan Africa and South Asia, are particularly vulnerable due to a combination of limited water infrastructure, overexploitation of available resources, and the impacts of climate change. Countries in these regions often experience seasonal water shortages, forcing communities to rely on unreliable and unsafe sources, which increases the risk of waterborne diseases.

Freshwater Ecosystems: Rivers, Lakes, Wetlands

Freshwater ecosystems—rivers, lakes, and wetlands—are essential to life on Earth. They provide water, support biodiversity, regulate the water cycle, and offer vital ecosystem services like food, transportation, and flood control. Understanding the characteristics and functions of each of these ecosystems is crucial for conserving them in the face of human activities and climate change.

Rivers

Rivers are dynamic systems that flow from higher elevations, such as mountains, to lower areas like oceans, seas, or lakes. Along their journey, rivers interact with a variety of landscapes, shaping the environment and supporting numerous species. The constant movement of water in rivers helps distribute nutrients, sediments, and organic matter, making them an important conduit for materials that sustain downstream ecosystems.

Rivers are categorized into different sections based on their flow, sediment load, and speed. **Headwaters** are the source of rivers, often found in mountainous regions where precipitation accumulates in small streams or springs. These upper reaches are typically fast-flowing and clear, with rocky beds that provide habitats for species like trout and insects adapted to cold, oxygen-rich waters. As rivers move downstream, they widen and slow down, allowing more sediments to settle. In these middle sections, species diversity increases, with fish, amphibians, and a variety of aquatic plants flourishing. The **lower reaches** of rivers, near their mouths, often support large floodplains that provide rich habitats for birds, mammals, and insects.

Rivers have a critical function in **shaping landscapes** by eroding land in one place and depositing sediments in another. This process creates fertile river valleys, making them ideal for agriculture. However, rivers are also vulnerable to human activities, such as damming, pollution, and water extraction, which can disrupt their natural flow and harm the species that depend on them.

Lakes

Lakes are standing bodies of freshwater that form in depressions in the Earth's surface. They range in size from small ponds to massive lakes like the **Great Lakes** in North America. Unlike rivers, lakes do not flow continuously, but they still have a key role in the water cycle, acting as reservoirs that store water, recharge groundwater supplies, and support diverse ecosystems.

Lakes are classified based on their nutrient levels and productivity. **Oligotrophic lakes** are nutrient-poor and have clear waters with low biological productivity. These lakes, often found in mountainous or cold regions, are home to fish species like trout and some aquatic plants. In contrast, **eutrophic lakes** are nutrient-rich and support a high level of biological activity, including abundant algae, fish, and aquatic plants. Eutrophication, however, can become a problem when excess nutrients, often from agricultural runoff or wastewater, lead to algae blooms that deplete oxygen and harm aquatic life.

Stratification is an important process in lakes, where layers of water with different temperatures form. In temperate regions, lakes undergo seasonal stratification. In the summer, warm water forms a layer on top of cooler, denser water, with a distinct boundary called the **thermocline**. This layering prevents mixing, which can limit the distribution of oxygen and nutrients in deeper layers. In the fall, cooling surface water sinks, causing the lake to mix and redistribute nutrients throughout the water column. This **turnover** is crucial for the health of lake ecosystems, allowing nutrients from the bottom to reach the surface and oxygen from the surface to reach the depths.

Lakes are important for **recreation, drinking water, and fisheries**, but they are vulnerable to pollution, invasive species, and climate change. Changes in temperature, precipitation, and human activities can alter water levels and water quality, impacting the species and ecosystems that depend on lakes.

Wetlands

Wetlands are unique ecosystems where water covers the soil or is present near the surface for part or all of the year. Wetlands can be **marshes, swamps, bogs, or fens**, depending on their hydrology, vegetation, and location. These ecosystems are incredibly diverse and serve as critical habitats for a wide range of species, including amphibians, birds, fish, and plants. Wetlands act as **biological filters**, trapping sediments and pollutants, improving water quality, and recharging groundwater supplies.

Wetlands provide several important ecosystem services, including **flood control** and **carbon sequestration**. During heavy rains or snowmelt, wetlands absorb excess water, preventing floods and reducing damage to nearby areas. Wetlands also store large amounts of carbon in their soil and vegetation, helping to mitigate climate change. For example, peatlands, a type of wetland, contain more carbon than all of the world's forests combined.

Coastal wetlands, such as mangroves and salt marshes, protect shorelines from erosion and provide nurseries for marine species like fish and shellfish. Inland wetlands, such as freshwater marshes and swamps, offer habitats for migratory birds, reptiles, and mammals. Despite their importance, wetlands are one of the most threatened ecosystems globally. Many wetlands have been drained or filled for agriculture, urban development, or infrastructure projects, leading to habitat loss and declines in biodiversity.

Wetlands are also vital for local economies, supporting activities like fishing, hunting, and tourism. Conservation efforts focus on protecting existing wetlands, restoring degraded ones, and ensuring that they continue to provide essential services. Protecting wetlands is also critical for maintaining the health of **river systems** and **lakes**, as they often act as buffers, absorbing excess nutrients and pollutants before they reach other water bodies.

Water Pollution: Causes and Solutions

Water pollution occurs when harmful substances, such as chemicals, waste, and pathogens, enter water bodies, making them unsafe for humans and wildlife. The causes of water pollution are varied and complex, but the solutions focus on reducing contaminants at the source and improving water treatment systems.

One of the primary causes of water pollution is **agricultural runoff**, which occurs when rain or irrigation water washes fertilizers, pesticides, and animal waste from farms into rivers, lakes, and oceans. These pollutants introduce excess nutrients like nitrogen and phosphorus into water bodies, leading to **eutrophication**. This process causes algae blooms, which can deplete oxygen levels in the water, creating "dead zones" where aquatic life cannot survive. Dead zones, such as those in the Gulf of Mexico, are a direct result of nutrient runoff from agriculture.

To address this, farmers can adopt more sustainable practices like **buffer strips** of vegetation between fields and waterways to absorb excess nutrients and prevent them from reaching water bodies. **Precision agriculture**—using technology to apply fertilizers and water more efficiently—can also reduce runoff.

Industrial pollution is another major cause of water contamination. Factories and power plants discharge pollutants, including heavy metals, chemicals, and radioactive waste, into rivers and oceans. In some cases, these discharges are the result of inadequate wastewater treatment. For example, **mercury contamination** from coal-fired power plants can accumulate in fish, posing health risks to humans who consume them. Governments and industries must enforce stricter regulations on wastewater treatment and reduce the release of harmful substances into water systems.

Oil spills represent another form of water pollution, often with devastating effects on marine ecosystems. When oil spills into the ocean, it forms a slick that suffocates marine life, damages coral reefs, and contaminates coastlines. While large-scale oil spills from tankers capture headlines, smaller, more frequent spills from ships and offshore drilling operations also contribute to long-term damage. **Stronger regulations on shipping** and **investments in safer drilling technologies** can help reduce the risk of oil spills.

Plastic pollution is a growing concern, especially in the world's oceans. Single-use plastics, such as bottles, bags, and packaging, break down into smaller particles known as **microplastics**, which can be ingested by fish and other marine animals. These plastics not only harm wildlife but can also enter the human food chain. Reducing plastic pollution requires a multi-faceted approach, including reducing plastic production, improving waste management systems, and promoting the use of biodegradable materials. Public awareness campaigns and government policies, such as bans on single-use plastics, are also essential in addressing the plastic pollution crisis.

Urban areas are another source of water pollution due to **untreated sewage** and stormwater runoff. In many developing countries, untreated sewage flows directly into rivers and lakes, contaminating water sources with pathogens that cause diseases like cholera and dysentery. In urban areas, stormwater runoff collects pollutants from streets, such as oil, heavy metals, and trash, and carries them into nearby water bodies. Improving **sanitation infrastructure** and installing **green infrastructure**, like permeable pavements and rain gardens, can help manage urban water pollution.

Finally, **climate change** is exacerbating water pollution problems. Increased rainfall and flooding can overwhelm sewage systems, leading to overflows of untreated waste into rivers and oceans. Warmer temperatures can also accelerate the growth of harmful algal blooms, further depleting oxygen in water bodies and creating more dead zones. Addressing climate change through **reducing greenhouse gas emissions** and investing in **resilient water infrastructure** is critical for mitigating the long-term impacts of water pollution.

Water Rights and International Conflicts

Water rights refer to the legal entitlements to use water from a source such as a river, lake, or groundwater. As freshwater resources become scarcer due to climate change, population growth, and increasing demand, water rights have become a critical issue, both within nations and across borders. In many parts of the world, the distribution of water rights leads to **international conflicts**, especially in regions where water sources are shared by multiple countries. These conflicts arise from disputes over how much water each nation is entitled to, who controls access, and how the resource should be managed.

One of the main drivers of water-related conflicts is the **uneven distribution of water** across the globe. While some regions have abundant freshwater resources, others struggle with scarcity. Rivers that cross borders, known as **transboundary rivers**, are often the source of tension. For example, the **Nile River** flows through eleven countries, including Egypt, Sudan, and Ethiopia, each of which depends on the river for water. Egypt, which is downstream, has historically controlled much of the water from the Nile, but Ethiopia's construction of the **Grand Ethiopian Renaissance Dam** has caused friction as Ethiopia seeks to use more of the river's water for its own development. Egypt fears that the dam will reduce the flow of water downstream, threatening its agricultural sector and overall water security.

International water conflicts are further complicated by legal issues related to **water rights**. In many cases, water rights are based on outdated agreements that no longer reflect current realities, such as changes in population, development, and environmental conditions. The 1959 Nile Waters Agreement between Egypt and Sudan, for example, allocated nearly all of the Nile's water to these two countries, with no provisions for upstream nations like Ethiopia. As the demand for water grows in the upstream countries, these old agreements become points of contention.

The **Jordan River Basin** is another example of a region where water rights are a source of international conflict. This river is shared by Israel, Jordan, Syria, Lebanon, and Palestine, and it provides water for agriculture, drinking, and industry. Water is a highly sensitive issue in the Middle East, where many countries face severe water shortages. Israel and Jordan, for instance, have long had disputes over access to the Jordan River's waters. However, the two countries signed a **water-sharing agreement** as part of their 1994 peace treaty, which helped reduce tensions over the resource. Despite such agreements, water remains a contentious issue in the region, particularly with Palestine, where water access is limited and tied to broader political conflicts.

Another challenge in managing shared water resources is that upstream countries often have an advantage over downstream countries. Upstream nations can build dams, divert water, or pollute the source before it reaches downstream users. This creates a power imbalance that can lead to conflict. For instance, the **Tigris and Euphrates Rivers**, which flow through Turkey, Syria, and Iraq, have been a source of tension for decades. Turkey, as the upstream country, has built several large dams as part of its **Southeastern Anatolia Project**, reducing the flow of water to Syria and Iraq. This has led to periodic disputes, particularly during times of drought when water becomes even more critical for agriculture and drinking supplies in the downstream countries.

Climate change is exacerbating these conflicts by altering **precipitation patterns** and reducing water availability in many regions. As glaciers melt and rainfall becomes less predictable, rivers and lakes that once provided reliable water supplies are shrinking. This creates further competition for the remaining water and increases the risk of conflict, particularly in already arid regions. In areas like the **Himalayan river basins**, where countries like India, China, Nepal, and Bangladesh rely on shared water resources, climate change is making water management more challenging, as the timing and amount of water flow become less predictable.

Efforts to resolve international water conflicts often involve **negotiation** and **cooperation**. International bodies like the **United Nations** and the **World Bank** have had a role in mediating disputes and facilitating agreements between nations. One approach is the development of **joint water management systems**, where countries that share a river or lake collaborate on managing the resource. The **Indus Waters Treaty** between India and Pakistan is often cited as a successful example of this approach. Despite long-standing political tensions, the two countries have cooperated over water-sharing since the treaty was signed in 1960, largely avoiding major conflicts over the Indus River.

However, not all conflicts are resolved through cooperation. In some cases, countries may take unilateral action, such as building dams or diverting rivers, leading to increased tension. When diplomatic solutions fail, disputes over water can escalate into broader geopolitical conflicts, especially in regions where water is scarce and populations are heavily dependent on a limited water supply.

CHAPTER 9: FORESTS AND GRASSLANDS

The Ecological Importance of Forests

Forests are vital ecosystems that support a vast array of biodiversity and provide essential services to the planet. Covering about 31% of the Earth's land surface, forests range from tropical rainforests to temperate and boreal forests, each offering unique ecological functions. The importance of forests goes beyond their beauty and biodiversity; they are critical to the global environment, human life, and the balance of natural systems.

One of the primary ecological roles of forests is their ability to act as **carbon sinks**. Trees absorb carbon dioxide (CO_2) from the atmosphere during photosynthesis, storing it as carbon in their trunks, branches, leaves, and roots. This process helps regulate the Earth's climate by removing a significant amount of CO_2, a major greenhouse gas that contributes to global warming. Tropical rainforests, such as the Amazon, are particularly important for carbon sequestration due to their dense vegetation and year-round growing season. However, deforestation and forest degradation release this stored carbon back into the atmosphere, exacerbating climate change.

Forests also contribute to **climate regulation** through their influence on local and global weather patterns. Forests, particularly tropical ones, generate humidity by releasing water vapor into the atmosphere through a process called **transpiration**. This water vapor contributes to cloud formation and precipitation, making forests critical to maintaining the water cycle. The loss of forests in one region can lead to reduced rainfall and increased droughts both locally and in distant regions. For example, the destruction of the Amazon rainforest is linked to reduced rainfall in parts of South America, which impacts agriculture and water supplies.

The **biodiversity** within forests is staggering. They are home to over 80% of terrestrial species, including mammals, birds, insects, and plants. Tropical rainforests, in particular, are known for their high species richness. Many of these species are endemic, meaning they are found nowhere else on Earth. The diverse structure of forests, with their multiple layers from the canopy to the forest floor, provides a variety of habitats for different species, allowing them to thrive. Forests also serve as important **gene pools**, providing genetic diversity that is essential for adaptation and resilience in changing environments. This diversity is vital not just for the health of the ecosystems but also for agricultural research, medicine, and biotechnology, where new plant compounds and genes are continually being discovered.

In addition to their role in biodiversity, forests are crucial for **water regulation**. Forests act like natural sponges, absorbing rainwater and slowly releasing it into rivers and streams. This regulates the flow of water, reducing the risk of floods during heavy rainfall and maintaining water supplies during dry periods. Forested watersheds supply clean drinking water to millions of people worldwide. By filtering water through soil and vegetation, forests reduce sedimentation and remove pollutants, ensuring higher water quality for ecosystems and human use. For example, the forests in the Pacific Northwest of the United States are important in providing clean water to urban areas like Seattle and Portland.

Soil conservation is another key ecological function of forests. Tree roots bind the soil together, reducing erosion caused by wind and water. Forest canopies also protect the soil from the direct impact of rainfall, which can otherwise wash away topsoil, the most fertile layer essential for plant growth. In tropical areas, where heavy rains are common, the presence of forests helps prevent landslides and maintains the fertility of the soil, which is crucial for agriculture and natural plant regeneration. Without forest cover, soil can degrade quickly, leading to desertification in extreme cases.

Forests have an essential role in **maintaining ecosystem services** that humans depend on for survival. These services include pollination, nutrient cycling, and the provision of timber, food, and medicinal plants. Indigenous communities around the world have relied on forests for thousands of years for food, shelter, and medicine. The knowledge these communities have about forest ecosystems is invaluable for sustainable management and conservation efforts. For instance, plants used in traditional medicines have led to the discovery of modern pharmaceuticals, highlighting the importance of preserving forest biodiversity.

Forests also provide habitats for species that are critical to **pollination**, such as bees, birds, and bats. Pollinators are essential for the reproduction of many plants, including crops that humans rely on for food. The loss of forest habitats can reduce pollinator populations, threatening food security in many regions. Similarly, forests are involved in **nutrient cycling**, where decomposing plant material enriches the soil with nutrients that support new plant growth, creating a self-sustaining system.

Forests also serve as **buffers against extreme weather events**. Mangrove forests, for example, protect coastal areas from the impacts of storms and tsunamis by reducing the force of waves and preventing coastal erosion. Similarly, forested areas can help mitigate the effects of storms and floods inland by slowing water runoff and providing natural barriers. This buffering capacity is becoming increasingly important as climate change leads to more frequent and severe weather events.

In addition to their ecological functions, forests have significant **cultural and recreational value**. Many indigenous cultures around the world see forests as sacred spaces and sources of spiritual nourishment. Forests also offer opportunities for recreation and tourism, providing economic benefits through activities like hiking, birdwatching, and eco-tourism. Protecting forests, therefore, not only safeguards biodiversity and ecosystem services but also preserves cultural heritage and supports local economies.

Deforestation and Reforestation Efforts

Deforestation, the clearing or thinning of forests by humans, has been a significant driver of environmental degradation. It is often driven by activities such as agriculture, logging, urban development, and mining. Deforestation has devastating consequences for ecosystems, climate regulation, and biodiversity. However, efforts to reverse these effects through reforestation—planting trees in deforested areas— are increasingly becoming a focus of global environmental conservation strategies.

One of the primary causes of deforestation is **agriculture**. As global populations grow, so does the demand for land to grow food. This often leads to the clearing of forests to make way for crops or livestock. The **slash-and-burn** technique, commonly used in tropical regions, involves cutting down trees and burning the land to clear it for agriculture. This not only destroys forests but also releases large amounts of carbon dioxide into the atmosphere, contributing to global warming. Deforestation for crops like soy, palm oil, and cattle ranching is particularly common in tropical areas like the Amazon rainforest, which is home to vast biodiversity.

Logging for timber is another major driver of deforestation. Forests are harvested for wood products, including furniture, paper, and construction materials. In many cases, logging is carried out unsustainably, with little regard for the long-term health of the forest. **Illegal logging** is also a problem in many countries, where forests are cut down without proper regulations or oversight, leading to rapid deforestation.

Deforestation has far-reaching impacts on the **global climate**. Trees act as carbon sinks, absorbing carbon dioxide from the atmosphere and storing it. When forests are cleared, not only is this carbon storage capacity lost, but the carbon stored in the trees is released back into the atmosphere, contributing to the greenhouse effect. Deforestation is responsible for nearly **10% of global greenhouse gas emissions**, making it a significant factor in climate change. Furthermore, the loss of forests disrupts local and global **water cycles**, as forests have a key role in generating rainfall through transpiration. In areas where deforestation occurs, local climates often become drier, leading to droughts and reduced agricultural productivity.

In response to the growing threat of deforestation, **reforestation** efforts have been gaining momentum. Reforestation involves planting trees in areas where forests have been cut down, with the aim of restoring ecosystems and the services they provide. Countries, NGOs, and international organizations have been working to scale up reforestation projects to offset the loss of forests. For example, **China's Great Green Wall project** is one of the largest reforestation efforts in the world. Launched in the 1970s, it aims to plant trees along the northern edge of the country to combat desertification and soil erosion, while also capturing carbon and improving local climates.

Agroforestry is another approach to reforestation that combines agriculture and tree planting. By integrating trees into farmland, agroforestry can improve soil health, increase biodiversity, and provide farmers with additional income from timber, fruits, or other tree products. This method not only helps restore forests but also makes farming more sustainable by enhancing the land's productivity and resilience to climate change.

Forest restoration projects are being implemented around the world to restore ecosystems and improve biodiversity. In Brazil, efforts to reforest parts of the **Atlantic Forest**, one of the most diverse and threatened ecosystems, are showing promise. Local communities and organizations are working to replant native species, restore degraded lands, and reconnect fragmented habitats to create corridors for wildlife. Similarly, in parts of Africa, the **Great Green Wall initiative** is a project that spans multiple countries and aims to combat desertification by planting trees across the Sahel region, helping to restore ecosystems and improve livelihoods for communities in the region.

Reforestation is not without its challenges. It can take decades for new forests to grow and begin providing the same ecosystem services as mature forests. Moreover, **monoculture plantations**, where only a single species of tree is planted, are often used in commercial reforestation efforts. These plantations do not provide the same biodiversity or ecological benefits as natural forests. Sustainable reforestation efforts focus on planting a diversity of native species to better mimic natural forest ecosystems.

Overall, while deforestation remains a major global issue, reforestation and restoration efforts are critical in addressing climate change, protecting biodiversity, and ensuring the sustainability of natural resources for future generations.

Grassland Ecosystems and Their Conservation

Grasslands are vast open areas where grasses dominate the landscape, providing critical habitats for a wide range of species and essential services to human populations. Found on every continent except Antarctica, grassland ecosystems include temperate prairies, savannas, and tropical grasslands. These ecosystems are under threat from human activities such as agriculture, urban development, and climate change, making their conservation increasingly important.

Biodiversity in grasslands is incredibly rich. Grasslands support a wide variety of wildlife, including large herbivores like bison, antelope, zebras, and elephants, as well as predators like lions, wolves, and cheetahs. Many grassland species are highly specialized, having evolved to survive in environments with periodic droughts, fires, and grazing pressure. For example, grasses in these ecosystems are adapted to being grazed and regrow quickly after being eaten by herbivores. This cycle of grazing and regrowth helps maintain the balance of grassland ecosystems, preventing any one species from dominating.

One of the most important ecological functions of grasslands is **carbon sequestration**. Although forests are often highlighted for their role in storing carbon, grasslands also are important. The roots of grasses extend deep into the soil, where they store carbon underground. Grasslands, particularly **temperate prairies**, have thick, fertile soils rich in organic matter, making them important carbon sinks. However, when grasslands are converted into cropland or overgrazed, the soil can become degraded, releasing stored carbon into the atmosphere and contributing to climate change.

Grasslands also have a critical role in **soil conservation** and **water regulation**. The deep root systems of grasses help anchor the soil, reducing erosion and maintaining soil health. In many regions, grasslands act as natural sponges, absorbing rainwater and slowly releasing it into streams and rivers, which helps regulate water flow and prevent flooding. Without healthy grasslands, these functions are lost, leading to increased soil erosion, water scarcity, and degraded landscapes.

One of the biggest threats to grasslands is **agricultural expansion**. Grasslands are often converted into cropland for growing crops like wheat, corn, and soybeans, or they are used for intensive livestock grazing. This conversion disrupts the natural balance of the ecosystem, reduces biodiversity, and leads to soil degradation. Overgrazing, where livestock eat grasses faster than they can regrow, can lead to desertification, where the land becomes barren and unable to support life.

Urbanization is another factor contributing to grassland loss. As cities expand, natural grasslands are often replaced by roads, buildings, and other infrastructure. This not only destroys the habitat for many species but also fragments the remaining ecosystems, making it difficult for wildlife to migrate, find food, and reproduce.

Conservation efforts to protect grasslands focus on **sustainable land management** and the creation of protected areas. In many regions, establishing **national parks** and wildlife reserves has helped protect large areas of grassland from agricultural development and urbanization. For example, **Serengeti National Park** in Tanzania and the **Great Plains** in North America are well-known for their efforts to preserve these unique ecosystems. In addition to protection, **restoration projects** aim to rehabilitate degraded grasslands by reintroducing native plant species and managing grazing more sustainably.

Grazing management is an important tool in grassland conservation. Sustainable grazing practices, such as **rotational grazing**, allow livestock to graze in one area for a short time before being moved to another area, giving grasses time to recover. This practice helps maintain the health of the grassland ecosystem and prevents overgrazing, which can lead to soil erosion and loss of biodiversity. In some cases, reintroducing native herbivores, such as bison in North America, helps restore the natural grazing patterns that are essential for grassland health.

Another promising approach is **community-based conservation**, where local communities are involved in the management and protection of grasslands. In many parts of Africa, community-managed conservancies allow local people to benefit from sustainable grazing and eco-tourism while protecting wildlife and maintaining the health of the grassland ecosystem.

Grasslands are often overlooked in conservation efforts, but their ecological importance cannot be understated. Protecting these ecosystems is critical not only for preserving biodiversity but also for maintaining the ecological services they provide, from carbon storage to water regulation. Sustainable land management, combined with protected areas and community involvement, offers a way forward in conserving grassland ecosystems for future generations.

Forests as Carbon Sinks and Climate Mitigation

Forests are often referred to as the "lungs of the Earth" because of their vital role in regulating the atmosphere and maintaining the balance of carbon dioxide (CO_2). As carbon sinks, forests absorb more carbon from the atmosphere than they release, making them essential in mitigating climate change. Understanding how forests function as carbon sinks and their role in climate mitigation is critical to addressing the global climate crisis.

The Carbon Cycle and Forests' Role

The carbon cycle describes the continuous movement of carbon through the Earth's atmosphere, oceans, soil, and living organisms. Carbon exists in various forms, including atmospheric CO_2, which is a key greenhouse gas contributing to global warming. Trees and forests are important in the carbon cycle through **photosynthesis**, a process by which trees absorb CO_2 from the air and use sunlight to convert it into glucose and oxygen. The carbon absorbed by trees is stored in their biomass—in trunks, branches, leaves, and roots—as well as in the soil.

Forests act as massive **carbon sinks**, meaning they capture and store carbon over long periods. This process reduces the concentration of CO_2 in the atmosphere, helping to regulate the global climate. Forests, especially old-growth forests and tropical rainforests, are particularly effective carbon sinks due to their dense biomass and high rates of photosynthesis. For example, the Amazon rainforest, often called the "world's largest carbon sink," absorbs billions of tons of CO_2 annually, making it a critical buffer against climate change.

Forests and Carbon Storage

Carbon is stored in forests in two main forms: **aboveground biomass** and **belowground biomass**. Aboveground biomass refers to the carbon stored in the trunks, branches, and leaves of trees. This is the most visible form of carbon storage and is relatively easy to measure. **Belowground biomass,** on the other hand, includes the carbon stored in tree roots and the soil. While less visible, this form of storage is incredibly important, as forests' soils can store even more carbon than the trees themselves.

Soil carbon sequestration is a long-term process. When trees die and decompose, much of their carbon is released back into the atmosphere, but some of it is absorbed into the soil, where it can remain locked away for hundreds or even thousands of years. Forest ecosystems, particularly in temperate and boreal regions, are especially adept at storing carbon in the soil. Boreal forests, which cover large parts of Canada, Russia, and Scandinavia, are estimated to store more than 30% of the world's soil carbon.

The ability of forests to store carbon is not constant. Forests in different regions, with varying types of vegetation and climate conditions, sequester carbon at different rates. Tropical forests, which are rich in biodiversity and experience year-round growth, tend to absorb more carbon compared to boreal forests, where growth is slower due to colder climates. However, boreal forests have the advantage of storing carbon for longer periods due to slower decomposition rates in cold environments.

Deforestation and Its Impact on Carbon Sinks

While forests serve as carbon sinks, **deforestation** transforms them into carbon sources. When forests are cleared for agriculture, logging, or urban development, the carbon stored in trees is released back into the atmosphere as CO_2, contributing to global warming. Deforestation is responsible for approximately **10% of global greenhouse gas emissions**, making it one of the largest contributors to climate change.

The loss of forest cover also reduces the Earth's ability to absorb future CO_2 emissions, creating a negative feedback loop. As deforestation continues, the global carbon sink capacity diminishes, accelerating the effects of climate change. This is especially concerning in tropical regions, where deforestation rates are highest. The

Amazon rainforest, for example, has seen significant forest loss in recent decades due to agricultural expansion, particularly for soy production and cattle ranching. If deforestation in the Amazon continues, the forest could reach a tipping point, where it no longer functions as a carbon sink but becomes a net carbon emitter.

Reforestation and Afforestation as Climate Solutions

In response to the negative impacts of deforestation, **reforestation** and **afforestation** are being promoted as key strategies for climate mitigation. Reforestation involves replanting trees in areas where forests have been degraded or destroyed, while afforestation involves planting trees in areas that were not previously forested.

Reforestation projects aim to restore the carbon sequestration capacity of forests, while also providing additional benefits such as restoring biodiversity, preventing soil erosion, and improving water quality. In many countries, large-scale reforestation efforts are underway. For instance, China's **Great Green Wall** project has resulted in the planting of millions of trees along the country's northern regions to combat desertification and reduce CO_2 emissions. Similarly, **Brazil's Atlantic Forest Restoration Pact** seeks to restore millions of hectares of degraded land, helping to absorb carbon and restore biodiversity.

Afforestation projects, such as **India's Green India Mission**, aim to expand forest cover and increase carbon sequestration. While afforestation can help mitigate climate change, it is essential to consider the type of trees planted. Planting monoculture plantations, where only one species of tree is planted, can lead to biodiversity loss and reduced ecosystem services. To maximize the benefits of afforestation, projects should focus on planting native species and creating diverse, resilient ecosystems.

Forest Management and Conservation

Sustainable forest management practices are crucial to maintaining forests' role as carbon sinks. **Selective logging**, for instance, can reduce the environmental impact of timber harvesting by only removing certain trees while leaving the overall structure of the forest intact. This practice allows the forest to continue sequestering carbon while providing timber resources. Additionally, **agroforestry**, which integrates trees into agricultural systems, can enhance carbon sequestration while improving crop yields and soil health.

Forest conservation is another critical aspect of climate mitigation. Protecting existing forests, particularly old-growth forests, is one of the most effective ways to maintain carbon sinks. These forests have accumulated carbon over centuries, and their loss would result in a significant release of CO_2. Conservation efforts include establishing **protected areas**, enforcing anti-logging laws, and engaging local communities in forest stewardship. For example, many indigenous communities have long practiced sustainable forest management, and their involvement in conservation efforts has proven successful in protecting forest ecosystems.

The Role of Forests in Global Climate Goals

Forests are integral to achieving global climate goals, including those set by the **Paris Agreement**. Under this agreement, countries have committed to reducing greenhouse gas emissions and limiting global temperature rise. Many nations have included forest conservation, reforestation, and afforestation in their national climate action plans, known as **Nationally Determined Contributions (NDCs)**.

Efforts like the **Bonn Challenge**, which aims to restore 350 million hectares of degraded and deforested land by 2030, are part of a broader global movement to enhance forests' role in climate mitigation. Forests not only absorb carbon but also provide critical services that support climate resilience, such as regulating water cycles, preventing floods, and protecting biodiversity.

CHAPTER 10: MARINE ECOSYSTEMS

Oceans: Key Features and Zonation

As mentioned, oceans cover about 71% of Earth's surface and hold around 97% of its water, making them one of the most prominent features of the planet. They are vast, dynamic systems that regulate climate, support marine life, and influence global weather patterns. Oceans are not uniform; they vary significantly in terms of depth, light availability, temperature, and nutrient concentration, leading to the development of distinct zones, each with its own ecological characteristics.

Key Features of Oceans

One of the most important features of oceans is their ability to **store and distribute heat**. Water has a high heat capacity, which means it can absorb a significant amount of heat without a large increase in temperature. This capacity allows oceans to moderate global climate by absorbing excess heat from the atmosphere and redistributing it through ocean currents. These currents, such as the **Gulf Stream** in the Atlantic Ocean, transport warm water from the equator toward the poles, while cold water moves from the poles toward the equator. This circulation helps regulate temperatures across the planet, making oceans essential to the climate system.

Oceans are also rich in **biodiversity**, home to millions of species ranging from microscopic plankton to the largest animal on Earth, the blue whale. Marine ecosystems, including coral reefs, kelp forests, and deep-sea environments, provide critical habitats for diverse species. Oceans support food chains that include everything from primary producers like phytoplankton to apex predators like sharks. Additionally, oceans serve as a major source of oxygen. Through **photosynthesis**, marine plants and algae, especially phytoplankton, produce at least 50% of the world's oxygen.

Zonation in Oceans

The ocean can be divided into distinct zones based on factors like light penetration, depth, and proximity to land. These zones—**pelagic, benthic, photic, aphotic, and intertidal**—each have unique conditions that influence the types of organisms that can survive there.

1.   Pelagic Zone

The **pelagic zone** is the open water region of the ocean, away from the shore and the ocean floor. It is further subdivided into several zones based on depth:

- **Epipelagic Zone**: Also known as the **sunlight zone**, this is the uppermost layer, extending to about 200 meters. It receives ample sunlight, allowing photosynthesis to occur. Phytoplankton, which forms the base of the marine food web, thrives here, supporting a wide range of marine life, including fish, sea turtles, and marine mammals.
- **Mesopelagic Zone**: The **twilight zone**, between 200 and 1,000 meters deep, receives very little sunlight, and no photosynthesis occurs here. However, many organisms, such as squid and bioluminescent fish, live in this zone. These creatures have adapted to the low-light conditions by developing special adaptations like large eyes or the ability to produce light through **bioluminescence**.
- **Bathypelagic Zone**: Known as the **midnight zone**, it stretches from 1,000 to 4,000 meters. This zone is completely dark, and temperatures are just above freezing. The animals that inhabit this zone, like giant squid and deep-sea anglerfish, rely on detritus falling from above, known as **marine snow**, or on predation of other deep-sea creatures.
- **Abyssopelagic and Hadalpelagic Zones**: The **abyss** (4,000 to 6,000 meters) and the **hadal zone** (deeper than 6,000 meters) represent the ocean's deepest parts. The pressure is extreme, and temperatures are near freezing. Despite these harsh conditions, unique organisms such as deep-sea worms, shrimp, and certain fish thrive around **hydrothermal vents**, which release heat and nutrients from the Earth's crust.

2.   Benthic Zone

The **benthic zone** refers to the ocean floor, from shallow coastal areas to the deepest parts of the ocean. The organisms that live here, called **benthos**, can vary widely depending on depth and substrate type. In shallow areas, benthic ecosystems include coral reefs, kelp forests, and seagrass meadows, all of which support high biodiversity. In deeper regions, the benthic zone is often dominated by scavengers and organisms adapted to high pressure and low temperatures. **Deep-sea vents** and **cold seeps** also support unique communities that rely on chemosynthesis rather than photosynthesis for energy.

3.   Photic Zone

The **photic zone** is the layer of the ocean that receives sufficient sunlight for photosynthesis, typically extending to depths of about 200 meters. This is where most of the ocean's primary production occurs, fueled by sunlight and nutrients. The organisms in this zone, such as **phytoplankton**, form the base of the marine food chain. Larger animals like fish, whales, and sea birds are common here, feeding on the abundance of life.

4. Aphotic Zone

Below the photic zone is the **aphotic zone**, which receives little to no sunlight. Photosynthesis cannot occur in this region, so organisms must rely on other sources of energy, such as **organic matter** falling from the surface or chemosynthetic processes at hydrothermal vents. Life in the aphotic zone is highly specialized, with many species exhibiting bioluminescence or other adaptations to survive in the dark, high-pressure environment.

5. Intertidal Zone

The **intertidal zone**, also known as the **littoral zone**, is the area where the ocean meets the land, between high and low tide lines. This zone experiences extreme fluctuations in conditions, including exposure to air, waves, and changing temperatures. Despite these challenges, it is rich in biodiversity, home to organisms like crabs, barnacles, sea stars, and various species of seaweed. Life in the intertidal zone must adapt to both submerged and exposed conditions, making it one of the most dynamic ecosystems in the ocean.

Overall, oceans are highly complex ecosystems with distinct zones that vary in depth, light availability, and pressure. Each zone—from the sunlit surface waters to the darkest ocean depths—supports a wide range of organisms uniquely adapted to the conditions they face. Understanding the key features of the ocean and its zonation is essential to appreciating its role in supporting life on Earth and regulating the planet's climate.

Coral Reefs and Mangroves

Coral reefs and mangroves are two of the most important coastal ecosystems, each providing critical services to marine life and coastal communities. Despite their differences in structure and location, these ecosystems are interconnected and play complementary roles in maintaining the health and stability of coastal environments.

Coral Reefs are found in warm, shallow waters, primarily in tropical and subtropical regions. They are built by **coral polyps**, tiny marine animals that secrete calcium carbonate, which forms the hard structures of reefs. Coral reefs are often referred to as the "rainforests of the sea" because of their incredible biodiversity. They provide habitat for about 25% of all marine species, including fish, mollusks, crustaceans, and many other organisms. Coral reefs support marine life by offering shelter, breeding grounds, and abundant food sources, making them one of the most productive ecosystems on Earth.

Beyond biodiversity, coral reefs provide critical **ecosystem services** to humans. They act as natural barriers, protecting coastlines from storm surges, waves, and erosion. This is particularly important in regions that are vulnerable to hurricanes and typhoons. Reefs also contribute significantly to local economies by supporting **fisheries** and **tourism**, drawing millions of visitors annually who engage in activities like snorkeling and diving.

However, coral reefs are highly sensitive to environmental changes. **Coral bleaching**, caused by rising ocean temperatures, occurs when corals expel the symbiotic algae living in their tissues, leading to loss of color and, eventually, coral death. Ocean acidification, pollution, and overfishing also pose serious threats to the health of coral reefs.

Mangroves, on the other hand, are coastal forests that grow in tropical and subtropical intertidal zones. Unlike coral reefs, which are underwater ecosystems, mangroves grow at the interface between land and sea, where they experience fluctuating water levels due to tides. Mangroves are characterized by their unique root systems, which are adapted to saltwater and provide stability in loose, muddy soils.

Mangroves provide numerous **ecosystem services** similar to coral reefs. Their dense root systems trap sediment, reducing coastal erosion and improving water quality. They act as **natural buffers**, protecting coastal communities from storm surges and tsunamis. Mangroves also serve as nurseries for many marine species, offering shelter and protection for young fish, crustaceans, and mollusks before they venture into the open sea.

Together, coral reefs and mangroves form interconnected coastal ecosystems that are vital for biodiversity, coastal protection, and the livelihoods of millions of people. Conservation efforts focused on both ecosystems are essential for sustaining the services they provide in the face of environmental challenges like climate change and human activity.

Marine Biodiversity and Ecosystem Services

Marine biodiversity refers to the variety of life found in the world's oceans, from the tiniest plankton to the largest whales. This diversity is critical for the functioning of marine ecosystems, as different species play specific roles in maintaining the health and stability of the ocean environment. The richness of marine biodiversity also provides essential **ecosystem services** that benefit human societies, from food and livelihoods to climate regulation and coastal protection.

Biodiversity in Marine Ecosystems

The oceans host a vast array of ecosystems, including coral reefs, mangroves, seagrass beds, and deep-sea habitats. Each of these ecosystems supports a distinct community of species, contributing to the overall diversity of marine life. For example, coral reefs are home to a high concentration of species, with thousands of species of fish, invertebrates, and algae living within these vibrant underwater structures. **Seagrass beds**, found in shallow waters, provide critical habitats for species like sea turtles, manatees, and fish. **Deep-sea ecosystems**, though less understood, harbor unique species adapted to the high-pressure, low-light conditions of the ocean's depths.

Biodiversity is essential for maintaining the balance of marine ecosystems. Each species occupies a specific **ecological niche**, contributing to processes like nutrient cycling, carbon storage, and the food web. **Keystone species**, such as sharks and sea otters, play particularly important roles in regulating populations of other species and maintaining ecosystem stability. For example, sharks help control the populations of smaller predatory fish, which in turn protects the health of coral reefs by preventing overgrazing of algae.

Ecosystem Services Provided by Marine Biodiversity

Marine ecosystems provide a wide range of services that are vital for human well-being. One of the most direct services is the provision of **food**. Oceans are a major source of protein for billions of people worldwide, with fisheries and aquaculture providing livelihoods for millions. Species like tuna, cod, and shrimp are critical to both local and global food security. Sustainable management of fisheries is essential to prevent overfishing and ensure the long-term availability of these resources.

Another key ecosystem service is **climate regulation**. Marine organisms, particularly phytoplankton and seagrasses, are important in the global carbon cycle. Phytoplankton, microscopic plants in the ocean, absorb large amounts of carbon dioxide during photosynthesis, helping to regulate the Earth's climate. Similarly, seagrass meadows and mangroves sequester carbon in their roots and soils, acting as important carbon sinks. Without these natural processes, atmospheric CO_2 levels would rise even more rapidly, exacerbating global warming.

Marine biodiversity also supports **coastal protection**. Ecosystems like coral reefs and mangroves act as natural barriers against storm surges, hurricanes, and tsunamis. By absorbing the energy of waves, these ecosystems reduce the impact of storms on coastal communities, preventing erosion and protecting infrastructure. The loss of these ecosystems would leave coastal areas more vulnerable to extreme weather events, which are becoming more frequent due to climate change.

In addition to food and coastal protection, marine ecosystems provide **cultural and recreational services**. Coastal and marine areas are popular destinations for tourism, contributing to the economies of many countries. Activities like snorkeling, scuba diving, and whale watching attract millions of visitors each year, creating jobs and generating income. The beauty and diversity of marine life also hold cultural significance for many coastal communities, particularly indigenous groups who have deep connections to the sea.

Threats to Marine Biodiversity

Despite their importance, marine ecosystems and biodiversity are under severe threat from human activities. **Overfishing**, habitat destruction, and **pollution** are some of the primary drivers of biodiversity loss in the oceans. **Illegal, unreported, and unregulated (IUU) fishing** depletes fish populations and disrupts the balance of marine food webs. Habitat destruction, such as the damage caused by bottom trawling and coastal development, reduces the availability of critical habitats for many species.

Climate change poses another significant threat to marine biodiversity. Warming ocean temperatures are causing coral bleaching, where coral polyps expel the algae that live within them, leading to the death of coral reefs. Ocean acidification, caused by the absorption of excess CO_2, weakens the shells of marine organisms like mollusks and corals, threatening entire ecosystems. Rising sea levels and more frequent extreme weather events also put additional pressure on marine environments and the species that depend on them.

Conservation and Sustainable Management

To protect marine biodiversity and the ecosystem services it provides, **conservation efforts** are essential. Establishing **marine protected areas (MPAs)**, where human activities are limited or restricted, is one of the most effective ways to conserve marine life. MPAs help preserve critical habitats, protect endangered species, and allow ecosystems to recover from human impacts. International cooperation is also crucial, as many marine species migrate across national boundaries.

In addition to MPAs, **sustainable fisheries management** is vital to prevent overexploitation of marine resources. Practices such as setting catch limits, regulating fishing gear, and protecting breeding grounds can help ensure that fish populations remain healthy. Public awareness and education campaigns about the importance of marine biodiversity and responsible consumption of seafood are also key components of a broader conservation strategy.

Deep Sea Ecosystems and Exploration

The deep sea is one of the most mysterious and least explored environments on Earth, covering over 60% of the planet's surface. It lies far below the ocean's surface, where sunlight cannot penetrate, temperatures are near freezing, and pressures are immense. Despite these extreme conditions, deep-sea ecosystems are teeming with life, hosting a wide variety of organisms adapted to survive in this harsh environment. As technology advances, scientists are uncovering the secrets of these ecosystems and gaining a deeper understanding of their importance to global biodiversity.

The Nature of Deep Sea Ecosystems

Deep-sea ecosystems exist in zones that start at about 200 meters below the surface, where sunlight fades, and extend down to the ocean floor, which can reach depths of 11,000 meters in areas like the **Mariana Trench**. Life in the deep sea is shaped by cold temperatures, high pressure, and the absence of light, creating an environment vastly different from shallow, sunlit waters.

Organisms in these ecosystems have developed remarkable adaptations. Many species in the **bathypelagic zone** (1,000 to 4,000 meters) rely on **bioluminescence** to communicate, hunt, and evade predators. Creatures like the **anglerfish** use light-producing organs to lure prey, while others, such as certain jellyfish, emit light as a defense mechanism. In the deeper **abyssal** and **hadal zones** (beyond 4,000 meters), animals like giant squid, deep-sea fish, and shrimp have adapted to survive in near-total darkness and crushing pressures.

Hydrothermal Vents and Chemosynthesis

One of the most fascinating deep-sea ecosystems is found around **hydrothermal vents**, which are cracks in the seafloor that release hot, mineral-rich water from beneath the Earth's crust. Discovered in 1977, hydrothermal vents host unique communities of organisms that rely on **chemosynthesis** instead of photosynthesis for energy. Chemosynthetic bacteria use the chemicals released from the vents, such as hydrogen sulfide, to produce energy, forming the base of the food web in these areas.

The animals that inhabit these ecosystems, including giant tubeworms, crabs, and deep-sea shrimp, have evolved to thrive in these extreme conditions. Tubeworms, for example, have no digestive system; instead, they rely on symbiotic bacteria within their bodies to convert chemicals from the vents into energy. Hydrothermal vent ecosystems demonstrate that life can exist in environments far removed from the sun's energy, reshaping our understanding of biological processes.

Exploration of the Deep Sea

Exploring the deep sea is a formidable challenge due to its depth, pressure, and lack of light. Until the development of **remotely operated vehicles (ROVs)** and **submersibles**, much of the deep ocean was inaccessible to humans. These technologies have revolutionized our ability to study deep-sea environments. Submersibles, like the famous **Alvin**, have allowed scientists to dive to extreme depths, while ROVs equipped with cameras and sampling tools can explore areas too deep for humans to visit directly.

Explorations have revealed not only the incredible biodiversity of the deep sea but also its potential for scientific discovery. Many species discovered in the deep ocean have unique biological compounds that could have medical or industrial applications. Furthermore, studying deep-sea ecosystems helps scientists understand how life on Earth can thrive in extreme environments, providing insights into the potential for life on other planets or moons.

Human Impact on Deep Sea Ecosystems

Despite their remoteness, deep-sea ecosystems are not immune to human activity. **Deep-sea mining**, driven by the demand for rare minerals used in electronics and renewable energy technologies, poses a significant threat. Mining the seafloor can disturb fragile ecosystems, destroy habitats, and release harmful sediments into the water. Similarly, **bottom trawling**, a fishing method that drags heavy nets along the ocean floor, can cause severe damage to deep-sea habitats, particularly coral reefs and sponge beds.

Pollution is another concern. Plastics, chemicals, and other pollutants from surface waters can make their way into the deep sea, where they accumulate and threaten marine life. Microplastics, in particular, have been found in deep-sea organisms, raising concerns about the long-term health of these ecosystems.

While exploration of the deep sea continues to reveal its ecological richness, it also highlights the need for conservation. Protecting these ecosystems from human exploitation and pollution is essential to preserve their biodiversity and the vital roles they play in global ecological processes.

Marine Pollution and Its Impact on Marine Life

Marine pollution is one of the most pressing environmental challenges facing the world today. Human activities have introduced various pollutants into the oceans, including plastics, chemicals, heavy metals, and oil, all of which have devastating impacts on marine ecosystems. These pollutants threaten the health of marine life, disrupt food chains, and compromise the services that oceans provide to humanity.

Types of Marine Pollution

One of the most visible and widespread forms of marine pollution is **plastic pollution**. Every year, millions of tons of plastic waste enter the oceans, much of it in the form of single-use plastics like bottles, bags, and packaging. Once in the ocean, plastics can take hundreds of years to decompose, breaking down into smaller particles known as **microplastics**. These microplastics are ingested by marine animals, from plankton to fish, and can cause physical harm or death. For example, sea turtles often mistake plastic bags for jellyfish, one of their primary food sources, leading to ingestion and intestinal blockages.

Chemical pollution is another significant threat to marine life. **Pesticides**, fertilizers, and industrial chemicals enter the oceans through agricultural runoff, rivers, and atmospheric deposition. These chemicals can have toxic effects on marine organisms. **Nutrient pollution**, particularly from nitrogen and phosphorus in fertilizers, leads to **eutrophication**, a process where excess nutrients cause harmful algae blooms. These blooms reduce oxygen levels in the water, creating **dead zones** where marine life cannot survive. The **Gulf of Mexico** is home to one of the largest dead zones, caused primarily by agricultural runoff from the Mississippi River Basin.

Oil spills represent another catastrophic form of marine pollution. When oil is released into the ocean, it spreads across the surface, forming a slick that blocks sunlight and prevents oxygen exchange. Oil-coated birds, fish, and marine mammals often die from hypothermia, poisoning, or suffocation. The **Deepwater Horizon oil spill** in 2010 is one of the most notorious examples, releasing millions of barrels of oil into the Gulf of Mexico and causing widespread environmental damage. The effects of this spill are still being felt today, with long-term impacts on marine life and coastal ecosystems.

Heavy metal pollution, particularly from mercury and lead, also poses serious risks to marine life. These metals enter the ocean through industrial waste and coal combustion, accumulating in the tissues of marine organisms. Mercury, for example, can build up in the bodies of fish and shellfish, posing a threat to predators higher up the food chain, including humans. **Bioaccumulation** and **biomagnification** are processes by which toxins like mercury become more concentrated as they move up the food chain, ultimately affecting apex predators such as sharks, tuna, and even marine mammals like dolphins and whales.

Impacts on Marine Life

Marine pollution affects organisms at every level of the food web. **Plankton**, the base of most marine food chains, is sensitive to changes in water quality. Pollution, particularly from chemicals and plastics, can reduce plankton populations, which in turn affects the species that rely on them for food. This disruption can cascade through the food web, impacting fish, marine mammals, and even seabirds.

Coral reefs, which are home to a diverse array of marine species, are particularly vulnerable to pollution. Chemicals, sediments, and plastics can smother corals, block sunlight, and hinder their ability to grow and reproduce. Coral bleaching, often exacerbated by pollution and warming waters, weakens coral ecosystems, making them more susceptible to disease and death.

Marine mammals, such as dolphins, whales, and seals, are among the most affected by marine pollution. Oil spills coat their bodies, damaging their insulating fur and causing hypothermia. Toxic chemicals and heavy metals accumulate in their bodies, leading to reproductive issues, immune system damage, and higher mortality rates. Many marine mammals are also threatened by **entanglement** in discarded fishing nets and plastic debris, which can cause injury, drowning, or starvation.

Sea turtles and seabirds are similarly at risk. Sea turtles often ingest plastic, mistaking it for food, which leads to digestive blockages and malnutrition. Seabirds, such as albatrosses, are frequently found with stomachs full of plastic waste, which prevents them from feeding properly and can lead to death.

Solutions to Marine Pollution

Addressing marine pollution requires a multifaceted approach, involving both **prevention** and **cleanup efforts**. Reducing plastic waste is a key component of prevention. Governments, NGOs, and industries are working to reduce the production and use of single-use plastics through bans, regulations, and public awareness campaigns. For example, countries like **Kenya** and **New Zealand** have implemented bans on plastic bags, significantly reducing the amount of plastic waste entering their oceans.

Improving waste management systems and promoting recycling can also help prevent plastic and other pollutants from reaching the ocean. In addition, developing **biodegradable alternatives** to plastics can reduce the longevity of plastic waste in marine environments.

Addressing chemical pollution requires stricter regulations on industrial waste disposal and agricultural runoff. Implementing **buffer zones** around agricultural fields and wetlands can help absorb excess nutrients before they reach water bodies, reducing the occurrence of harmful algal blooms. Similarly, international cooperation is necessary to reduce mercury emissions and other toxic chemicals from industrial sources. The **Minamata Convention on Mercury**, for example, is an international treaty aimed at reducing mercury pollution, which has already shown positive effects in reducing mercury levels in the environment.

Oil spill prevention and response measures are also critical. Strict regulations on oil drilling, transport, and extraction are essential to prevent spills from occurring in the first place. In the event of a spill, rapid response techniques, such as containment booms, skimmers, and chemical dispersants, can minimize environmental damage. Research into better spill-cleanup technologies, such as oil-eating bacteria, is ongoing and may offer more effective solutions in the future.

In terms of **marine habitat restoration**, organizations are working to rehabilitate damaged ecosystems. Coral restoration projects, for example, involve growing coral fragments in nurseries and transplanting them onto degraded reefs. Mangrove reforestation and seagrass restoration efforts are also helping to restore coastal ecosystems that have been harmed by pollution.

Public awareness and **education** are important in reducing marine pollution. Campaigns to educate people about the impact of their waste, promote recycling, and reduce plastic use can drive behavior change on an individual level. Coastal cleanups, organized by groups like the **Ocean Conservancy**, engage volunteers in removing plastic and other debris from shorelines, preventing it from reaching marine ecosystems.

International collaboration is vital to addressing marine pollution, as oceans do not adhere to national boundaries. Treaties like the **London Convention** on ocean dumping and the **MARPOL** Convention on marine pollution from ships have established global frameworks for reducing pollution at sea. Continued cooperation between nations, industries, and conservation organizations is essential to tackling the complex, global issue of marine pollution.

CHAPTER 11: AIR QUALITY AND POLLUTION

Major Air Pollutants: Sources and Impacts

Air pollutants come from various natural and human activities, and they have significant impacts on health, the environment, and climate. Understanding the sources and effects of these pollutants is essential for tackling air quality issues. There are several major air pollutants that are commonly monitored and regulated, each affecting the environment in different ways. These include particulate matter (PM), nitrogen oxides (NOx), sulfur dioxide (SO_2), carbon monoxide (CO), ground-level ozone (O_3), and volatile organic compounds (VOCs).

Particulate Matter (PM)

Particulate matter, or PM, consists of tiny particles suspended in the air. These particles can be either **PM10**, which are particles with diameters of 10 micrometers or smaller, or **PM2.5**, which are fine particles 2.5 micrometers or smaller. The smaller the particles, the more dangerous they are because they can penetrate deep into the lungs and even enter the bloodstream.

Sources of PM include **industrial emissions, vehicle exhaust**, and **construction activities**, which release dust and soot into the air. **Natural sources** of PM include wildfires, dust storms, and volcanic eruptions. **Fine particulate matter** is often produced by the combustion of fossil fuels, such as coal, oil, and gas, and is a major component of urban smog.

The health impacts of particulate matter are severe. Exposure to PM can lead to **respiratory diseases** like asthma, bronchitis, and chronic obstructive pulmonary disease (COPD). Long-term exposure can cause cardiovascular problems and increase the risk of **premature death**. The environmental impact includes reduced visibility (haze) and damage to ecosystems, as particles can settle on water bodies, soil, and plants, affecting their health.

Nitrogen Oxides (NOx)

Nitrogen oxides (NOx), primarily **nitric oxide (NO)** and **nitrogen dioxide (NO_2)**, are gases produced from high-temperature combustion. They are major pollutants emitted by **vehicles, power plants**, and **industrial facilities**. NOx forms when nitrogen in the air reacts with oxygen under high temperatures, such as in car engines or coal-burning power stations.

NOx gases contribute to the formation of **ground-level ozone** and **smog**, which are harmful to human health. They also lead to the formation of **acid rain**, which can damage forests, lakes, and buildings. NOx emissions are linked to respiratory issues, particularly in vulnerable populations like children and the elderly. Prolonged exposure to high levels of NOx can increase the risk of lung diseases and lower resistance to respiratory infections like the flu.

Sulfur Dioxide (SO_2)

Sulfur dioxide (SO_2) is a colorless gas with a sharp odor, produced mainly from the **burning of fossil fuels** containing sulfur, such as coal and oil. **Power plants, industrial processes**, and **refineries** are major sources of SO_2 emissions. **Volcanic activity** also releases significant amounts of sulfur dioxide into the atmosphere.

When sulfur dioxide is released into the air, it reacts with water vapor and other chemicals to form **sulfuric acid**, which contributes to **acid rain**. Acid rain damages forests, harms aquatic ecosystems by lowering the pH of water bodies, and accelerates the decay of buildings and monuments. SO_2 can also irritate the respiratory system, causing short-term symptoms like coughing and throat irritation. Chronic exposure can lead to severe lung conditions and heart disease.

Carbon Monoxide (CO)

Carbon monoxide (CO) is a colorless, odorless gas produced by the incomplete combustion of carbon-containing fuels such as **gasoline, natural gas**, and **wood**. The primary sources of CO include **vehicle emissions, residential heating**, and **industrial processes**. In urban areas, motor vehicles are the largest contributors to carbon monoxide pollution.

CO is particularly dangerous because it binds to hemoglobin in the blood more easily than oxygen, reducing the amount of oxygen that reaches the body's organs and tissues. Even low levels of exposure can cause headaches, dizziness, and nausea, while high levels can lead to impaired cognitive function, unconsciousness, and even death. People with heart conditions are particularly vulnerable because CO exposure can exacerbate symptoms of **angina** (chest pain) and other cardiovascular problems.

Ground-Level Ozone (O_3)

Ozone (O_3) in the **upper atmosphere** protects the Earth from the sun's harmful ultraviolet radiation. However, **ground-level ozone** is a harmful pollutant that forms when nitrogen oxides (NOx) and volatile organic compounds (VOCs) react in the presence of sunlight. This reaction typically occurs during hot, sunny days, making ozone pollution a significant issue in urban areas with heavy traffic and industrial activity.

Ozone at ground level is a major component of **smog** and poses serious health risks. It can cause **respiratory problems**, including chest pain, throat irritation, and difficulty breathing. Prolonged exposure to high ozone levels can aggravate asthma, reduce lung function, and increase susceptibility to respiratory infections. Ground-level ozone also damages vegetation, reducing agricultural productivity and affecting forest ecosystems.

Volatile Organic Compounds (VOCs)

Volatile organic compounds (VOCs) are a group of chemicals that easily evaporate into the air. They are emitted by a wide range of products and activities, including **vehicle exhaust, industrial processes, solvents**, and **paints. Natural sources** of VOCs include vegetation, particularly trees like pines and eucalyptus, which release organic compounds into the air.

VOCs are precursors to ground-level ozone formation and are involved in the creation of **photochemical smog.** Many VOCs, such as **benzene, formaldehyde,** and **toluene**, are toxic and can cause a variety of health problems, including **eye and respiratory tract irritation, headaches**, and **dizziness**. Long-term exposure to certain VOCs, such as benzene, is known to increase the risk of cancer.

The Impact of Major Air Pollutants on Climate

In addition to their effects on human health and ecosystems, major air pollutants also have a significant impact on the climate. Particulate matter, for example, can either **cool or warm** the atmosphere, depending on the type of particles. **Sulfate aerosols**, which come from sulfur dioxide, reflect sunlight and can have a cooling effect, whereas **black carbon**, or soot, absorbs sunlight and warms the atmosphere.

Ground-level ozone and VOCs, along with carbon monoxide, are **indirect greenhouse gases** because they contribute to the formation of ozone in the lower atmosphere, which acts as a **greenhouse gas**. Ozone traps heat, contributing to **global warming**. Nitrogen oxides also contribute to this process by enhancing ozone formation.

Efforts to Reduce Air Pollution

Efforts to reduce major air pollutants focus on **regulation, technological innovation**, and **public awareness**. Governments have enacted laws such as the **Clean Air Act** in the United States, which sets limits on the amount of pollutants that can be emitted by industries and vehicles. **Emission controls** on cars, such as catalytic converters, have significantly reduced carbon monoxide, NOx, and VOC emissions. In power plants, **scrubbers** are used to remove sulfur dioxide from emissions, while switching from coal to cleaner energy sources like **natural gas** or **renewable energy** further reduces pollutants.

In recent years, the promotion of **electric vehicles (EVs)** and renewable energy sources has gained momentum as a strategy to reduce emissions of carbon monoxide, NOx, and particulate matter. Public transportation, carpooling, and alternative fuels like hydrogen also have a role in reducing air pollution in cities.

Acid Rain: Causes and Effects

Acid rain refers to any form of precipitation—rain, snow, fog, or even dust—that is more acidic than normal. It occurs when sulfur dioxide (SO_2) and nitrogen oxides (NOx) are released into the atmosphere, where they react with water, oxygen, and other chemicals to form sulfuric and nitric acids. These acids then fall back to Earth, either in wet or dry forms, causing damage to the environment, human health, and infrastructure.

Causes of Acid Rain

The primary sources of acid rain are **fossil fuel combustion** and industrial processes. When fossil fuels like coal, oil, and natural gas are burned in power plants, factories, and vehicles, they release large quantities of sulfur dioxide and nitrogen oxides into the atmosphere. **Coal-fired power plants** are a major contributor to SO_2 emissions, while **motor vehicles** and industrial processes are significant sources of NOx emissions.

Once released into the atmosphere, SO_2 and NOx undergo a series of chemical reactions. These gases combine with **water vapor** and **oxygen** to form **sulfuric acid** (H_2SO_4) and **nitric acid** (HNO_3). These acids can remain suspended in the atmosphere for days or even weeks, allowing them to be carried by winds across long distances before falling back to the ground in the form of acid rain. This means that regions far from the sources of pollution can still be affected by acid rain.

Natural sources, such as volcanic eruptions and wildfires, also release sulfur dioxide and nitrogen oxides, but human activities are responsible for the vast majority of acid rain-causing emissions.

Effects of Acid Rain

Acid rain has widespread and destructive effects on the environment, particularly on **forests, soils, lakes,** and **streams**. One of the most visible impacts is on forests, where acid rain damages leaves and bark, making trees more vulnerable to diseases, insects, and harsh weather. In particular, **high-elevation forests**, such as those in the Appalachian Mountains, are especially affected because the clouds and fog at higher altitudes are often more acidic than rainfall. Acid rain can strip nutrients like calcium and magnesium from the soil, which are essential for tree health, weakening trees and slowing their growth.

Acid rain also affects **aquatic ecosystems**, particularly in regions with soils that have little buffering capacity to neutralize the acidity. Lakes and streams in these areas can become highly acidic, killing fish and other aquatic organisms. **Fish populations** are particularly sensitive to changes in pH. For example, in highly acidified lakes, species such as trout and salmon struggle to survive, as the acidic water disrupts their reproductive processes and can cause deformities in developing fish. In some extreme cases, entire aquatic ecosystems can collapse due to acidification, leading to a significant loss of biodiversity.

The effects of acid rain are not limited to natural ecosystems; they also extend to **human-built structures**. Acid rain accelerates the corrosion of metals and the deterioration of building materials, especially those made of limestone and marble, which are particularly susceptible to acid. Famous monuments, such as the **Taj Mahal** and **Acropolis**, have suffered damage from acid deposition over the years. Acid rain also affects **painted surfaces** and **vehicles**, leading to increased maintenance costs.

Acid rain has direct and indirect effects on **human health** as well. While acid rain itself does not pose a direct health risk, the pollutants that cause acid rain—sulfur dioxide and nitrogen oxides—are harmful. These gases contribute to **respiratory problems**, such as asthma and bronchitis, especially in children and the elderly. In urban areas, the formation of **fine particulate matter (PM2.5)** from SO_2 and NOx emissions can exacerbate these health issues, leading to premature deaths.

Mitigation Efforts

Efforts to reduce the impact of acid rain have focused on regulating emissions of SO_2 and NOx through policies such as the **Clean Air Act** in the United States and **European Union directives** on air quality. **Scrubbers** installed in power plants help remove sulfur dioxide from emissions, while catalytic converters in cars reduce nitrogen oxide emissions. Additionally, shifting away from coal and other fossil fuels to **renewable energy sources** like wind, solar, and natural gas has reduced the overall levels of pollutants that cause acid rain.

Ozone Depletion and UV Radiation

The depletion of the ozone layer, a protective shield of gas in the Earth's stratosphere, has raised significant concerns over the past few decades. Ozone (O_3) in the stratosphere absorbs much of the sun's harmful ultraviolet (UV) radiation, protecting living organisms from its damaging effects. However, human activities have contributed to the thinning of this protective layer, leading to an increase in UV radiation reaching the Earth's surface, which has significant environmental and health consequences.

Causes of Ozone Depletion

Ozone depletion is primarily caused by **chlorofluorocarbons (CFCs)** and other **halogenated ozone-depleting substances (ODS)**, such as **halons** and **carbon tetrachloride**. CFCs were widely used as refrigerants, aerosol propellants, and solvents because they were chemically stable, non-toxic, and non-flammable. However, it is this chemical stability that allows CFCs to remain in the atmosphere for decades, eventually reaching the stratosphere.

In the stratosphere, **ultraviolet radiation** breaks down CFC molecules, releasing **chlorine atoms**. These chlorine atoms act as catalysts in a series of reactions that destroy ozone molecules. A single chlorine atom can destroy thousands of ozone molecules before it is deactivated. **Bromine**, found in halons used in fire extinguishers, is even more destructive to ozone than chlorine. The loss of ozone in the stratosphere allows more UV-B radiation to reach the Earth's surface.

The most dramatic example of ozone depletion occurs over **Antarctica**, where a phenomenon known as the **ozone hole** forms each year during the Southern Hemisphere's spring (September to November). Cold temperatures in the polar stratosphere create **polar stratospheric clouds (PSCs)**, which provide surfaces for chlorine and bromine compounds to react and destroy ozone at an accelerated rate.

Effects of Ozone Depletion

The depletion of the ozone layer has significant implications for both human health and the environment. One of the most direct impacts is the increase in **UV-B radiation** that reaches the Earth's surface. Increased UV-B exposure can lead to a variety of health problems, particularly **skin cancer**. **Malignant melanoma**, the deadliest form of skin cancer, is strongly linked to UV exposure, as are **basal** and **squamous cell carcinomas**. People with lighter skin tones are especially vulnerable, but all populations are at increased risk with higher UV levels.

In addition to skin cancer, UV-B radiation can cause **cataracts**, a clouding of the eye's lens that leads to vision impairment and blindness. **Suppressed immune function** is another potential health effect, as UV radiation can weaken the immune system's ability to fight off certain diseases, making people more susceptible to infections and illnesses.

The environment also suffers from increased UV radiation. **Phytoplankton**, the foundation of marine food webs, are highly sensitive to UV-B radiation. In polar regions, where the ozone layer is most depleted, reductions in phytoplankton populations have been observed, which can have ripple effects throughout the entire marine ecosystem. This reduction in phytoplankton also impacts the global carbon cycle, as these organisms are important in absorbing carbon dioxide from the atmosphere.

UV-B radiation affects **terrestrial plants** as well, reducing their growth and productivity. Crops such as wheat, rice, and soybeans are particularly vulnerable to increased UV exposure, which can reduce crop yields and threaten food security. Additionally, UV-B radiation can degrade **materials** like plastics, wood, and fabrics, shortening their lifespan and increasing maintenance costs.

Mitigation and Recovery of the Ozone Layer

Global efforts to address ozone depletion have been highly successful, particularly through the adoption of the **Montreal Protocol** in 1987. This international treaty aimed to phase out the production and use of CFCs and other ozone-depleting substances. Thanks to the Montreal Protocol and its subsequent amendments, the use of CFCs has dramatically decreased, and the ozone layer is showing signs of recovery. Scientists predict that, with continued adherence to the treaty, the ozone layer will return to its pre-1980 levels by the middle of the 21st century.

The successful reduction in ozone-depleting substances is a powerful example of international cooperation and effective environmental regulation. However, the ozone layer remains fragile, and the continued use of some ozone-depleting substances in certain industries means that vigilance is still necessary to ensure its full recovery.

Indoor Air Pollution and Public Health

Indoor air pollution is a significant yet often overlooked public health concern. It refers to the presence of harmful pollutants inside buildings and homes, where people spend a large portion of their time. These pollutants can originate from a variety of sources, including household products, cooking, heating, and building materials. The health impacts of indoor air pollution can be severe, especially for vulnerable groups such as children, the elderly, and people with pre-existing respiratory conditions.

One of the main sources of indoor air pollution is the **use of solid fuels** such as wood, coal, and biomass for cooking and heating. In many developing countries, people rely on open fires or poorly ventilated stoves, which release large amounts of harmful pollutants like **carbon monoxide (CO), particulate matter (PM2.5),** and **volatile organic compounds (VOCs)**. These pollutants can lead to serious health problems, including **chronic obstructive pulmonary disease (COPD)**, **lung cancer**, and **pneumonia**, particularly in women and children who are often most exposed to indoor cooking smoke.

Tobacco smoke is another major source of indoor air pollution, contributing to a range of health issues. Secondhand smoke contains thousands of chemicals, many of which are known carcinogens. Exposure to tobacco smoke indoors can cause respiratory infections, worsen asthma, and increase the risk of heart disease and stroke.

Building materials and household products also contribute to indoor air pollution. **Asbestos**, once commonly used in construction, can release fibers that, when inhaled, cause lung diseases such as **asbestosis** and **mesothelioma**. Additionally, **lead-based paints** and other materials can release toxic particles into the air, posing risks of **lead poisoning**, especially in older buildings. **Formaldehyde**, found in pressed wood products and insulation, is another indoor pollutant linked to respiratory problems and cancer.

Household chemicals, including **cleaning products, pesticides**, and **air fresheners**, often contain harmful VOCs, which can lead to headaches, dizziness, and long-term respiratory issues. Poor **ventilation** exacerbates the problem by trapping these pollutants indoors, allowing them to accumulate to hazardous levels.

The health effects of indoor air pollution are wide-ranging, from **irritation of the eyes, nose, and throat** to more serious conditions like **heart disease** and **cancer**. Children, in particular, are at greater risk due to their developing lungs and higher breathing rates. Moreover, indoor air pollution is linked to increased rates of **premature death**, particularly in low-income households where solid fuels are commonly used.

Efforts to reduce indoor air pollution include improving **ventilation**, using **cleaner cooking fuels**, eliminating tobacco smoke indoors, and ensuring that building materials and household products are free from harmful substances. Raising public awareness and implementing policies to promote safer indoor environments are essential for protecting public health from the dangers of indoor air pollution.

CHAPTER 12: CLIMATE CHANGE AND ITS IMPACTS

The Science of Climate Change

Climate change refers to long-term changes in temperature, precipitation, and other atmospheric conditions on Earth. While the climate naturally fluctuates over centuries due to factors such as volcanic activity and variations in solar radiation, the current pattern of rapid warming is directly linked to human activities. Understanding the science behind climate change involves examining how human actions, particularly the burning of fossil fuels, have altered the Earth's energy balance and triggered changes in weather patterns and ecosystems.

At the heart of climate change is the **greenhouse effect**, a natural process that keeps the Earth warm enough to support life. The Earth's atmosphere contains gases like **carbon dioxide (CO_2), methane (CH_4)**, and **water vapor**, which trap some of the sun's heat and prevent it from escaping back into space. This process is essential for maintaining temperatures that allow life to thrive. Without the greenhouse effect, the Earth would be about 33°C colder, making it uninhabitable.

However, human activities have significantly increased the concentration of greenhouse gases in the atmosphere, enhancing the natural greenhouse effect. The burning of **fossil fuels** such as coal, oil, and natural gas releases large amounts of CO_2, the primary driver of global warming. **Deforestation**, which reduces the number of trees that can absorb CO_2, and **agriculture**, which releases methane from livestock and rice paddies, also contribute to the rise in greenhouse gases. Since the Industrial Revolution, atmospheric CO_2 levels have increased by more than 40%, from about 280 parts per million (ppm) to over 415 ppm today.

When more greenhouse gases are added to the atmosphere, they trap additional heat, causing the Earth's average temperature to rise. This warming is known as **global warming**, and it is a key driver of climate change. Since the late 19th century, global temperatures have risen by approximately 1.1°C. While this may seem small, even slight increases in global temperatures can have far-reaching impacts on weather patterns, ecosystems, and sea levels.

Climate models, which are mathematical simulations of the Earth's climate system, have a critical role in understanding how greenhouse gases affect the planet. These models take into account complex interactions between the atmosphere, oceans, land, and ice to predict future climate changes based on different scenarios of greenhouse gas emissions. Climate models have consistently shown that human activities are responsible for most of the warming observed over the past century. They also predict that if greenhouse gas emissions continue to rise at their current rate, global temperatures could increase by 2 to 4°C by the end of this century.

The warming of the Earth's surface has numerous effects on the climate system. One of the most significant is the **melting of polar ice**. Arctic sea ice, which reflects sunlight and helps keep the planet cool, is shrinking at an alarming rate. The loss of ice contributes to the warming of the polar regions and creates a feedback loop: as the ice melts, more dark ocean water is exposed, which absorbs more heat, accelerating the warming process. The melting of **glaciers** and **ice sheets** in Greenland and Antarctica contributes to **sea level rise**, which threatens coastal cities and low-lying islands.

Warmer temperatures also increase the amount of **water vapor** in the atmosphere, which is itself a greenhouse gas. This creates another feedback loop: as temperatures rise, more water evaporates, adding to the greenhouse effect and causing even more warming. Changes in temperature and moisture levels also affect **weather patterns**, leading to more frequent and intense extreme weather events like **heatwaves**, **hurricanes**, and **floods**. For instance, warmer ocean temperatures provide more energy for tropical storms, making them stronger and more destructive.

Another critical consequence of climate change is the **acidification of the oceans**. Oceans absorb about 30% of the CO_2 emitted by human activities, which reacts with seawater to form carbonic acid. This process lowers the pH of the ocean, making it more acidic. Ocean acidification harms marine life, particularly organisms like **coral**, **shellfish**, and **plankton** that rely on calcium carbonate to build their skeletons and shells. As acid levels rise, these organisms struggle to survive, which can disrupt entire marine food webs.

Ecosystems on land are also being affected by rising temperatures and changing precipitation patterns. Plants and animals that are adapted to specific climate conditions are forced to migrate to new areas, if possible, in order to survive. **Boreal forests**, which thrive in colder climates, are being replaced by temperate forests as temperatures rise, and **species extinctions** are becoming more common as habitats change faster than species can adapt. In many regions, **droughts** are becoming more frequent and severe, leading to water shortages and agricultural declines.

The **feedback mechanisms** involved in climate change make it a self-reinforcing process. For example, thawing **permafrost** in the Arctic releases methane, a potent greenhouse gas, which further accelerates warming. Similarly, the destruction of forests reduces the planet's ability to absorb CO_2, which allows more greenhouse gases to accumulate in the atmosphere.

Despite the complexity of the climate system, the basic science behind climate change is clear: human activities, especially the burning of fossil fuels, are causing the Earth to warm at an unprecedented rate. This warming is already having profound impacts on weather patterns, ecosystems, and human societies. Understanding the science behind these changes is crucial for developing effective strategies to mitigate further warming and adapt to the changes that are already underway.

Impacts on Ecosystems and Species

Climate change is dramatically reshaping ecosystems and threatening countless species around the world. As global temperatures rise, weather patterns shift, and sea levels increase, ecosystems must adapt to new conditions, often at a pace that many species cannot keep up with. These changes disrupt the delicate balance that has sustained life for millennia, leading to loss of biodiversity, altered food webs, and habitat destruction.

One of the most significant impacts of climate change on ecosystems is **shifts in species distribution**. As temperatures rise, many species are migrating to higher altitudes or latitudes in search of cooler climates. For example, fish populations in the North Atlantic and Pacific Oceans are moving toward the poles as ocean waters warm. Terrestrial species like butterflies and birds are also shifting their ranges, often finding new habitats in cooler areas. However, not all species can migrate. **Plants** and animals that are highly specialized to specific environments, such as alpine or polar species, have fewer options and may face extinction.

Coral reefs, one of the most biodiverse ecosystems on Earth, are particularly vulnerable to climate change. Warming ocean temperatures cause **coral bleaching**, a phenomenon where corals expel the symbiotic algae that provide them with food and give them their vibrant colors. Without these algae, corals can starve and die, leading to the collapse of the entire reef ecosystem. Coral reefs support a wide variety of marine life, and their decline has cascading effects on fish populations, coastal protection, and the livelihoods of communities that depend on fishing and tourism.

Polar ecosystems are undergoing some of the most rapid changes due to climate change. In the Arctic, sea ice is shrinking at unprecedented rates, reducing critical habitat for species like polar bears, seals, and walruses. These animals rely on sea ice for hunting and breeding, and as it disappears, they face significant challenges in finding food and raising their young. Similarly, **penguins** and other cold-adapted species in Antarctica are struggling to cope with melting ice shelves and changing prey availability.

Forests are also experiencing significant stress due to climate change. Warmer temperatures and changing precipitation patterns are increasing the frequency and severity of **wildfires**, which devastate forest ecosystems and release large amounts of stored carbon into the atmosphere. Additionally, **pest infestations**, such as bark beetles in North American forests, are becoming more widespread due to milder winters, weakening trees and making them more susceptible to disease.

Ocean acidification, another consequence of climate change, poses a severe threat to marine life. As oceans absorb more carbon dioxide, the water becomes more acidic, disrupting the ability of organisms like **shellfish**, **plankton**, and **corals** to form their shells and skeletons. These organisms form the foundation of marine food webs, and their decline could have far-reaching effects on entire ecosystems, from small fish to large marine mammals.

The impact of climate change on ecosystems is not just limited to the natural world —human populations are also affected. **Agriculture** is particularly vulnerable to changing weather patterns, with crop yields declining due to droughts, floods, and shifting growing seasons. This can lead to food insecurity, particularly in regions that are already prone to hunger and poverty.

The accelerating rate of species loss and ecosystem disruption is a warning that climate change's impacts are profound and far-reaching. The loss of biodiversity reduces ecosystems' resilience, making them more vulnerable to further changes and diminishing the services they provide, such as water purification, carbon storage, and pollination.

Climate Change Mitigation and Adaptation

Addressing climate change requires a combination of **mitigation**—reducing greenhouse gas emissions—and **adaptation**—adjusting to the changes that are already occurring. These two strategies are essential to slow the pace of climate change and protect communities and ecosystems from its most severe impacts.

Mitigation focuses on reducing the amount of carbon dioxide (CO_2) and other greenhouse gases released into the atmosphere. The primary sources of these emissions are the **burning of fossil fuels** for energy, deforestation, and industrial processes. One of the most effective ways to mitigate climate change is by transitioning to **renewable energy sources**, such as wind, solar, and hydropower. These technologies generate electricity without producing CO_2, helping to reduce the carbon footprint of the energy sector.

Another important mitigation strategy is **reforestation** and **forest conservation**. Trees absorb carbon dioxide as they grow, acting as natural carbon sinks. Protecting existing forests and planting new trees can help offset emissions from other sources. **Sustainable agricultural practices** also are influential in mitigation by improving soil carbon storage and reducing methane emissions from livestock.

In the transportation sector, the shift to **electric vehicles (EVs)**, public transit, and improved fuel efficiency can significantly reduce emissions from one of the largest sources of pollution. Governments around the world are increasingly setting targets for **carbon neutrality**, aiming to achieve a balance between emissions produced and emissions removed from the atmosphere by the middle of the century.

While mitigation is critical to preventing further warming, some level of climate change is now inevitable, making **adaptation** equally important. Adaptation involves making changes to social, economic, and environmental systems to minimize the damage caused by rising temperatures, extreme weather, and other climate impacts.

In coastal areas, where **sea-level rise** and storm surges pose significant threats, adaptation measures include the construction of **sea walls**, **mangrove restoration**, and the development of **flood-resistant infrastructure**. These measures help protect communities from flooding and erosion while preserving ecosystems that provide natural barriers against storms.

Agriculture must also adapt to changing conditions. Farmers are adopting **drought-resistant crops**, improving **irrigation efficiency**, and adjusting planting schedules to cope with altered growing seasons. These practices help maintain food production in the face of more frequent and severe droughts, floods, and heatwaves.

Cities, too, are taking steps to adapt to climate change. Urban planners are incorporating **green spaces**, such as parks and green roofs, to reduce the urban heat island effect and improve air quality. Cities are also investing in **resilient infrastructure** that can withstand extreme weather events, such as hurricanes and floods, reducing the risk of damage and displacement.

International cooperation is essential for both mitigation and adaptation efforts. Initiatives like the **Paris Agreement** aim to limit global temperature rise by reducing emissions and providing support for countries that are most vulnerable to climate impacts. By combining mitigation and adaptation strategies, we can reduce the severity of climate change and build a more resilient future for both people and ecosystems.

Climate Change and Extreme Weather Events

Climate change is intensifying extreme weather events around the world. Rising global temperatures and shifting weather patterns are making events like hurricanes, floods, droughts, heatwaves, and wildfires more frequent and severe. The connection between climate change and extreme weather lies in the way a warming atmosphere affects the dynamics of weather systems, leading to more volatile conditions that can cause widespread damage to ecosystems, infrastructure, and human populations.

One of the most significant impacts of climate change is the increase in **heatwaves**. As global temperatures rise, heatwaves are becoming longer, more intense, and more frequent. These prolonged periods of extreme heat have direct consequences for human health, agriculture, and infrastructure. Heatwaves can cause dehydration, heatstroke, and exacerbate pre-existing health conditions, leading to higher mortality rates, especially among vulnerable populations like the elderly, children, and those with chronic illnesses. Urban areas, in particular, are susceptible to the **urban heat island effect**, where concrete and asphalt absorb and retain heat, making cities significantly hotter than surrounding rural areas.

Droughts are another extreme weather event that is being exacerbated by climate change. Warmer temperatures increase the rate of evaporation from soil and water bodies, leading to drier conditions, especially in regions already prone to arid climates. Prolonged droughts can devastate agriculture, reduce water supplies, and increase the risk of wildfires. In countries like Australia, the southwestern United States, and parts of Africa, droughts are becoming more intense, leading to crop failures, food insecurity, and economic hardship for communities dependent on agriculture. Additionally, the reduction in freshwater availability due to droughts can exacerbate water conflicts in regions where resources are already scarce.

At the opposite extreme, climate change is also leading to **increased precipitation** in some areas, resulting in more frequent and severe **floods**. A warmer atmosphere holds more moisture, which means that when storms do occur, they tend to produce more rain. This is particularly evident in tropical regions and areas affected by monsoons, where heavier downpours are becoming more common. Flash floods, which occur when intense rainfall overwhelms drainage systems, can cause significant damage to infrastructure, displace populations, and disrupt local economies. Coastal areas are particularly vulnerable to **storm surges**—a rise in sea level caused by intense storms—compounded by rising sea levels due to climate change.

Hurricanes and **tropical storms** are another type of extreme weather event that is being influenced by climate change. While the overall frequency of hurricanes may not increase, there is strong evidence that the intensity of these storms is rising. Warmer ocean temperatures provide more energy to fuel storms, leading to hurricanes with higher wind speeds, more rainfall, and greater potential for destruction. The **Atlantic hurricane season** has seen several record-breaking storms in recent years, such as Hurricane Maria in 2017 and Hurricane Dorian in 2019, both of which caused catastrophic damage. These stronger storms not only threaten coastal communities but also leave long-lasting impacts on economies, public health, and infrastructure.

Wildfires are also becoming more frequent and intense as a result of climate change. Higher temperatures and prolonged droughts create ideal conditions for wildfires to ignite and spread rapidly. Regions like **California**, **Australia**, and parts of southern Europe have experienced some of their most destructive wildfire seasons in recent years. Wildfires not only destroy homes, forests, and wildlife but also release large amounts of carbon dioxide into the atmosphere, further contributing to global warming. Smoke from wildfires also degrades air quality, posing serious health risks to people living downwind from affected areas.

In addition to directly influencing weather events, climate change also alters the behavior of large-scale atmospheric patterns like the **jet stream**, which can lead to prolonged periods of extreme weather. For example, a slower, meandering jet stream can cause weather systems to linger over an area for longer than usual, leading to extended heatwaves, droughts, or storms. This phenomenon was observed during the **European heatwave** of 2019 and the **Texas winter storm** of 2021, where unusual jet stream patterns contributed to the persistence of extreme conditions.

The increased frequency and intensity of extreme weather events due to climate change present significant challenges for communities around the world. Adapting to these changes requires improving **infrastructure resilience**, enhancing **disaster preparedness**, and investing in **early warning systems**. For example, building **flood defenses**, retrofitting buildings to withstand stronger storms, and implementing **water management strategies** to cope with droughts are critical steps in reducing the impacts of extreme weather.

CHAPTER 13: ENVIRONMENTAL HEALTH AND TOXICOLOGY

Toxins in the Environment: Types and Sources

Toxins in the environment refer to harmful substances that can cause damage to living organisms, including humans, plants, and animals. These toxins can originate from natural sources, but human activities have introduced a wide range of synthetic chemicals that pose significant risks to environmental and public health. Understanding the types and sources of these toxins is crucial for managing their impact and reducing exposure.

Types of Environmental Toxins

There are several categories of toxins commonly found in the environment, each with distinct chemical properties and health effects.

1. **Heavy Metals:** Heavy metals such as **lead, mercury, arsenic**, and **cadmium** are naturally occurring elements that can become toxic in high concentrations. They are particularly concerning because they do not degrade in the environment and can accumulate in living organisms over time, a process known as **bioaccumulation**. Lead, for example, is found in contaminated soils and old paint, and it can cause neurological damage, especially in children. Mercury, often released from coal-burning power plants, accumulates in fish and can lead to mercury poisoning when consumed by humans.

2. **Persistent Organic Pollutants (POPs):** POPs are chemicals that remain in the environment for long periods and can travel long distances through air and water. These include substances like **polychlorinated biphenyls (PCBs), dioxins**, and **DDT**, a pesticide once widely used but now banned in many countries due to its environmental and health risks. POPs are particularly dangerous because they are **lipophilic**, meaning they accumulate in the fatty tissues of animals and humans, leading to long-term exposure and increased risk of cancer, reproductive issues, and immune system damage.

3. **Pesticides and Herbicides:** Chemicals used in agriculture to control pests and weeds, such as **glyphosate, atrazine**, and **organophosphates**, are significant sources of environmental toxins. While they help increase crop yields, they also pose risks to wildlife and human health. Pesticides can contaminate water supplies through agricultural runoff, and prolonged exposure has been linked to various health problems, including hormone disruption, cancer, and developmental disorders.

4. **Endocrine Disruptors:** Endocrine disruptors are chemicals that interfere with the body's hormonal systems. Common endocrine disruptors include **bisphenol A (BPA)**, found in plastics, and **phthalates**, used in personal

care products and industrial materials. These chemicals mimic or block hormones like estrogen and testosterone, leading to reproductive issues, developmental problems, and increased risk of diseases like cancer.

5. **Airborne Toxins**: Many air pollutants act as toxins, with significant sources including industrial emissions, vehicle exhaust, and burning fossil fuels. **Volatile organic compounds (VOCs)**, such as benzene and formaldehyde, are commonly found in urban air due to vehicle emissions and industrial activities. These toxins contribute to **smog formation** and pose direct risks to human health, causing respiratory problems, cardiovascular disease, and cancer.

Sources of Environmental Toxins

Toxins enter the environment through both natural processes and human activities. Understanding the main sources of these pollutants helps identify ways to reduce exposure and mitigate their effects.

1. **Industrial Processes**: Many heavy metals and POPs are byproducts of industrial activities. Factories release toxic chemicals through emissions into the air, discharges into water, or direct dumping onto land. For example, chemical plants may release **PCBs** and **dioxins** during the production of plastics and other materials, while mining activities often result in the release of arsenic and cadmium into soil and water.

2. **Agriculture**: Pesticides and herbicides are a major source of environmental toxins. **Agricultural runoff** carries these chemicals into rivers, lakes, and groundwater, where they can contaminate drinking water and harm aquatic ecosystems. Fertilizers, often used in excess, can also contribute to nutrient pollution, leading to toxic algal blooms that release harmful toxins into water bodies.

3. **Transportation**: Vehicle exhaust is a key source of airborne toxins like nitrogen oxides (NOx), VOCs, and particulate matter (PM2.5). These pollutants contribute to **urban air pollution** and can trigger respiratory issues and cardiovascular problems in exposed populations. In cities with heavy traffic, air quality often suffers due to the high concentration of these toxins.

4. **Waste Disposal**: Improper disposal of hazardous waste, including **e-waste** (discarded electronics), can introduce toxins like lead, mercury, and cadmium into the environment. Landfills that are not properly managed may leak toxins into nearby soil and water, affecting local ecosystems and human populations. In developing countries, where waste management practices are often inadequate, the problem is particularly severe.

5. **Consumer Products**: Everyday products such as plastics, cleaning agents, and personal care products can also be sources of environmental toxins. For instance, plastics often contain **BPA** and **phthalates**, which can leach into the environment during production, use, or disposal. These chemicals then enter waterways or are released into the air, contributing to long-term environmental contamination.

6. **Natural Sources**: While human activities are the primary contributors to environmental toxins, some originate from natural processes. **Volcanic eruptions** release large amounts of sulfur dioxide and heavy metals into the atmosphere, and **wildfires** can release toxic gases and particulate matter. Although these natural events contribute to environmental pollution, they are typically less persistent than human-induced sources.

Understanding the types and sources of environmental toxins is essential for developing effective strategies to reduce exposure and mitigate the harmful effects of these pollutants. Reducing industrial emissions, improving waste management, and promoting the use of safer chemicals in consumer products are key steps in addressing this global issue.

Bioaccumulation and Biomagnification

Bioaccumulation and biomagnification are processes through which harmful substances, such as toxins and pollutants, concentrate in organisms and move up through food chains, leading to increasing levels of contamination in higher trophic levels.

Bioaccumulation refers to the gradual build-up of substances like heavy metals, pesticides, and other pollutants in an organism over time. This happens when an organism absorbs toxins faster than it can excrete or metabolize them. Toxins, particularly those that are fat-soluble, can accumulate in the fatty tissues of organisms, leading to long-term exposure. For example, fish in contaminated waters may accumulate mercury or PCBs (polychlorinated biphenyls) in their tissues. Since these substances do not break down easily, the toxins stay within the organism and build up over time, often causing harmful effects on the organism's health, such as reproductive issues, developmental problems, or even death.

Biomagnification, on the other hand, occurs when the concentration of a toxin increases as it moves up the food chain. This process starts with lower organisms, like plankton, that absorb small amounts of toxins from their environment. When predators, such as small fish, consume these contaminated organisms, they ingest higher levels of the toxin. As larger predators, like larger fish or birds, eat smaller contaminated organisms, the concentration of toxins magnifies with each step up the food chain. A classic example of biomagnification is the case of **DDT**, a pesticide that accumulated in birds of prey like eagles and falcons. These birds consumed fish that had ingested smaller, contaminated organisms, leading to high DDT concentrations, which caused eggshell thinning and population declines.

Heavy metals like mercury are also subject to biomagnification. In aquatic environments, mercury enters the water through industrial pollution and is converted to **methylmercury**, a highly toxic form. Small organisms in the water absorb methylmercury, which is then passed up the food chain to larger fish and marine mammals. Humans who consume these fish, such as tuna or swordfish, are exposed to dangerous levels of mercury, which can affect the nervous system and lead to cognitive impairments, particularly in developing fetuses.

Bioaccumulation and biomagnification are significant concerns for both **ecosystems and human health**. Top predators, including humans, are most vulnerable to the effects of biomagnification because they receive the highest doses of contaminants. The persistence of these toxins in the environment and their ability to accumulate in living organisms make them a long-term hazard, requiring strict monitoring and regulation to reduce exposure.

Human Health and Environmental Hazards

Human health is closely linked to the environment, and exposure to environmental hazards can lead to a wide range of health problems. These hazards include air pollution, contaminated water, toxic chemicals, and unsafe waste disposal practices. Each of these hazards poses risks that can affect the quality of life and longevity.

Air pollution is one of the most pervasive environmental hazards. Pollutants like **particulate matter (PM2.5), nitrogen oxides (NOx),** and **volatile organic compounds (VOCs)** can lead to respiratory illnesses, heart disease, and stroke. Long-term exposure to poor air quality, particularly in urban and industrial areas, increases the risk of chronic conditions such as asthma, bronchitis, and lung cancer.

Contaminated water is another significant hazard. Exposure to pollutants such as **heavy metals** (e.g., lead, mercury), **pesticides**, and **industrial chemicals** can cause severe health issues. Lead in drinking water, for example, is known to cause neurological damage, particularly in children, leading to developmental delays and cognitive impairments. Waterborne pathogens, such as **E. coli** and **Salmonella**, also pose risks, causing gastrointestinal illnesses and, in severe cases, death.

Chemical exposure from household products, industrial processes, and pesticides can lead to both acute and chronic health problems. For example, exposure to **pesticides** can disrupt the nervous system, lead to reproductive issues, and increase the risk of cancer. Long-term exposure to **solvents** or **cleaning products** can affect the liver, kidneys, and immune system.

Waste disposal practices, particularly in areas without proper waste management systems, expose populations to toxic materials like **e-waste**, which contains hazardous substances like **lead, cadmium,** and **mercury**. Improper handling and disposal can lead to soil and water contamination, impacting both human health and local ecosystems.

Understanding and managing environmental hazards is essential for protecting human health, especially in communities where exposure to these hazards is high. Preventive measures, such as improving air quality, ensuring safe drinking water, and regulating chemical usage, are key strategies for reducing the health impacts of environmental hazards.

Endocrine Disruptors and Long-Term Health Impacts

Endocrine disruptors are chemicals that interfere with the endocrine system, which regulates hormones in the body. These chemicals can mimic, block, or alter hormonal signals, leading to long-term health impacts that affect development, reproduction, metabolism, and even behavior.

One of the most common endocrine disruptors is **bisphenol A (BPA)**, a chemical found in plastics and food packaging. BPA can mimic estrogen, a hormone critical to reproductive health, and has been linked to **infertility, developmental issues**, and **breast cancer**. Studies have shown that even low levels of BPA exposure, especially during critical periods of development such as fetal growth and early childhood, can have long-lasting effects on health. **Phthalates**, another group of endocrine disruptors found in personal care products and plastics, have been associated with reduced fertility, birth defects, and altered development in children.

Pesticides, such as **DDT** and **atrazine**, are also known endocrine disruptors. DDT, although banned in many countries, persists in the environment and can still be found in human tissues. It has been linked to **reproductive issues**, particularly in women, leading to reduced fertility and increased risk of miscarriage. Atrazine, a commonly used herbicide, has been shown to affect the reproductive systems of amphibians and may pose similar risks to humans, including hormone imbalances and developmental defects.

Endocrine disruptors can also have significant effects on the **thyroid gland**, which regulates metabolism, growth, and brain development. **Polychlorinated biphenyls (PCBs)**, industrial chemicals used in electrical equipment and banned in many countries, are linked to thyroid hormone disruptions. These disruptions can lead to developmental delays, reduced IQ, and cognitive impairments, particularly in children exposed in utero or during early life.

Long-term exposure to endocrine disruptors may also increase the risk of **metabolic disorders** such as obesity and diabetes. Some disruptors, such as **perfluoroalkyl substances (PFAS)**, used in non-stick cookware and waterproof clothing, have been associated with altered metabolism and insulin resistance, which can contribute to the development of **type 2 diabetes**.

The cumulative and long-term effects of endocrine disruptors raise concerns about their impact on public health. Many of these chemicals persist in the environment for long periods, leading to continuous low-level exposure over time. Reducing exposure to endocrine disruptors through stricter regulations on chemicals in consumer products, improved waste management, and public awareness is critical for mitigating their harmful health impacts.

CHAPTER 14: SUSTAINABLE ENERGY

Fossil Fuels: Impacts and Alternatives

Fossil fuels, including **coal, oil**, and **natural gas**, have powered industrial growth and modern society for over a century. They provide the energy for electricity generation, transportation, and heating. However, the environmental and health impacts of relying on fossil fuels are severe, making it essential to explore alternative, cleaner energy sources.

One of the most significant impacts of fossil fuels is their contribution to **climate change**. When burned, coal, oil, and natural gas release large amounts of **carbon dioxide (CO_2)**, a major greenhouse gas. CO_2 traps heat in the Earth's atmosphere, leading to global warming. Since the Industrial Revolution, CO_2 concentrations have risen from around 280 parts per million (ppm) to over 415 ppm today, primarily due to fossil fuel combustion. This increase in greenhouse gases has led to more frequent and intense **heatwaves, droughts**, and **extreme weather events**, which are directly linked to the changing climate.

In addition to CO_2, fossil fuels emit other harmful pollutants. **Sulfur dioxide (SO_2)**, primarily from coal, contributes to **acid rain**, which damages forests, soils, and freshwater ecosystems. **Nitrogen oxides (NOx)**, released from vehicles and power plants, lead to the formation of **ground-level ozone**, or smog, which causes respiratory issues and aggravates conditions like asthma. Particulate matter from coal and oil combustion is also a major health hazard, contributing to **lung disease, heart problems**, and **premature death**.

Fossil fuel extraction and production have their own environmental consequences. **Coal mining** leads to habitat destruction, water contamination, and in the case of **mountaintop removal**, irreversible damage to landscapes. **Oil drilling** can result in spills that devastate marine ecosystems, like the 2010 **Deepwater Horizon** disaster in the Gulf of Mexico. **Hydraulic fracturing (fracking)**, used to extract natural gas, risks contaminating groundwater with chemicals and methane, contributing to both local pollution and global climate change.

Given the many downsides of fossil fuels, the transition to **alternative energy sources** is essential for reducing their environmental impact. **Renewable energy** offers a sustainable and clean solution. **Solar power**, harnessing energy from the sun, has become one of the fastest-growing energy sources. Solar panels convert sunlight directly into electricity without producing any emissions. With technological advances and decreasing costs, solar energy is becoming more accessible for homes, businesses, and even large-scale power generation.

Wind energy is another promising alternative. Wind turbines capture kinetic energy from the wind and convert it into electricity. Like solar, wind energy is emissions-free once the infrastructure is in place, and wind farms can generate significant amounts of electricity, particularly in coastal and plains regions. Offshore wind farms are growing rapidly, taking advantage of stronger and more consistent winds at sea.

Hydropower, which generates electricity by harnessing the energy of flowing water, is already widely used, especially in large dam projects. While hydropower is a renewable energy source, it can have environmental impacts such as disrupting fish migration and altering ecosystems downstream of dams. Efforts to improve the design of hydropower plants are addressing these challenges.

Geothermal energy taps into the Earth's internal heat, providing a consistent and reliable energy source. This energy is most viable in areas with geothermal activity, such as Iceland and parts of the United States. **Biomass energy**, derived from organic materials like crop waste, wood, and animal manure, is also a renewable alternative. While it emits CO_2 when burned, it can be part of a carbon-neutral cycle if managed sustainably, as the CO_2 released is roughly equivalent to what the plants absorbed during growth.

One of the key benefits of renewable energy is that it does not deplete natural resources or generate significant pollution during operation. Unlike fossil fuels, renewables are not subject to market fluctuations driven by scarcity, geopolitical tensions, or extraction difficulties. However, the shift to renewables requires investment in infrastructure and energy storage solutions to ensure consistent power supply, particularly as solar and wind are intermittent sources of energy.

In the long run, reducing reliance on fossil fuels through renewable energy adoption is essential for mitigating climate change, improving air quality, and ensuring a more sustainable future.

Renewable Energy: Solar, Wind, Hydro, Geothermal

Renewable energy sources are increasingly becoming the backbone of efforts to reduce greenhouse gas emissions and combat climate change. Unlike fossil fuels, which are finite and polluting, renewable energy sources such as **solar, wind, hydropower**, and **geothermal** are abundant, cleaner, and sustainable. These energy sources harness natural processes to generate power without depleting resources or significantly harming the environment.

Solar Energy is derived from the sun's radiation and converted into electricity using **solar photovoltaic (PV) panels**. Solar panels capture sunlight and transform it into electrical energy through the photovoltaic effect, a process where sunlight excites electrons in the panel's cells to generate an electric current. Solar energy is one of the most widely adopted renewable energy sources because of its versatility and declining costs. Solar panels can be installed on rooftops for individual homes or used in vast **solar farms** to generate power for entire communities. Advances in solar technology, including **concentrated solar power (CSP)**, allow for more efficient energy capture and storage, making it possible to generate electricity even when the sun is not shining.

Solar energy's main benefits include its zero-emission operation and the fact that sunlight is a virtually limitless resource. However, solar panels have some limitations, such as their dependence on weather conditions and daylight hours, meaning energy storage systems like **batteries** are needed to ensure a continuous power supply.

Wind Energy harnesses the kinetic energy of the wind to turn large turbines, which convert the motion into electrical energy. Wind farms are typically located in areas with consistent wind patterns, such as coastal regions or open plains. **Offshore wind farms** are gaining popularity because winds over the ocean are often stronger and more reliable than on land. Wind turbines are becoming more efficient and are now capable of generating significant amounts of electricity without emitting pollutants.

The advantages of wind energy include its relatively low cost and its minimal environmental impact once turbines are operational. The major challenges with wind energy are its intermittency—wind does not blow all the time—and the need for suitable locations, as turbines require vast open spaces or offshore locations. There are also concerns about **noise pollution** and impacts on wildlife, particularly birds and bats, though technological innovations are addressing some of these issues.

Hydropower generates electricity by using the energy of flowing water, typically from rivers or dams. Hydropower is one of the oldest and most widely used forms of renewable energy. Large hydropower plants use **dams** to store water in reservoirs, releasing it through turbines to generate electricity. Smaller **run-of-the-river** projects do not require dams and have a lower environmental impact, making them an attractive option for regions where large dam projects are impractical.

Hydropower is highly efficient and provides a consistent source of energy, as water flow is more predictable than solar or wind patterns. However, large hydropower projects can disrupt ecosystems by altering natural water flow, impacting fish populations, and flooding large areas. In some cases, the decay of organic matter in reservoirs can also release **methane**, a potent greenhouse gas, which offsets some of the environmental benefits of hydropower.

Geothermal Energy harnesses the Earth's internal heat, which originates from the planet's core and remains accessible in areas with volcanic or tectonic activity.

Geothermal plants convert this heat into electricity by using steam from hot water reservoirs deep underground to spin turbines. In addition to electricity generation, geothermal energy is also used for **direct heating** in homes and industries.

Geothermal energy is extremely reliable and available 24/7, as the Earth's internal heat is constant. It also has a minimal environmental footprint once operational. However, geothermal plants are limited to specific regions with access to geothermal reservoirs, such as Iceland, the western United States, and parts of Asia. Initial installation costs are high, but the long-term benefits make it a competitive renewable energy source.

Energy Efficiency and Conservation

Energy efficiency and conservation are critical strategies for reducing energy consumption, lowering greenhouse gas emissions, and building a sustainable energy future. While renewable energy sources like solar and wind provide cleaner alternatives, using energy more efficiently and conserving resources can reduce the overall demand for energy, making the transition to renewables more feasible and effective.

Energy efficiency refers to the practice of using technology that requires less energy to perform the same function. This approach focuses on reducing energy waste without sacrificing comfort or productivity. For example, **LED light bulbs** use a fraction of the electricity that traditional incandescent bulbs require to produce the same amount of light. Similarly, **energy-efficient appliances**, such as refrigerators, washing machines, and air conditioners, use less electricity while maintaining or even improving performance compared to older models.

One of the most significant areas for improving energy efficiency is in **buildings**. Homes and commercial buildings consume large amounts of energy for heating, cooling, and lighting. **Insulating buildings**, installing **double-glazed windows**, and using **smart thermostats** are effective ways to reduce energy consumption by maintaining temperature control without excessive heating or cooling. Moreover, **green building designs** that incorporate **passive solar heating**, natural ventilation, and energy-efficient materials can significantly reduce a building's energy footprint.

Transportation is another sector where energy efficiency can make a significant impact. **Electric vehicles (EVs)** are more energy-efficient than traditional gasoline-powered cars, converting a higher percentage of energy from the battery to power the wheels. Public transportation systems, such as electric buses or high-speed rail, are also more efficient ways to move people over long distances compared to individual car travel. In addition, **fuel-efficient technologies** like hybrid engines and improved aerodynamics can reduce fuel consumption in conventional vehicles.

Energy conservation, on the other hand, involves changing behaviors to reduce energy use. Unlike efficiency, which focuses on using less energy to perform the same task, conservation emphasizes reducing or eliminating unnecessary energy consumption altogether. Simple actions like turning off lights when not in use, **unplugging electronics**, and using **energy-saving settings** on devices can lead to significant reductions in energy demand. Conservation also involves lifestyle choices, such as walking or cycling instead of driving, reducing air travel, and using public transportation more frequently.

Industrial processes also offer vast potential for energy savings through both efficiency and conservation measures. **Manufacturing plants** can adopt more efficient machinery, improve **process heating**, and recover waste heat to reduce overall energy consumption. Optimizing **supply chains** to minimize transportation and waste can further decrease the energy needed for production.

One of the key benefits of energy efficiency and conservation is that they often provide **immediate cost savings**. Businesses and households that adopt energy-saving practices or invest in efficient technologies can lower their energy bills while reducing their environmental impact. Over time, these savings can offset the initial costs of upgrading to more efficient systems or appliances.

Governments and organizations are important in promoting energy efficiency through **policies**, **incentives**, and **education**. Programs like **Energy Star** certification help consumers identify energy-efficient products, while **tax credits** and **rebates** encourage the adoption of clean technologies. At a broader level, national energy policies that set **efficiency standards** for appliances, vehicles, and buildings are essential for driving large-scale improvements.

Decentralized Energy Systems and Microgrids

Decentralized energy systems, often referred to as **distributed energy resources (DERs)**, represent a shift from the traditional centralized model of power generation to a more localized, flexible system of energy production and distribution. These systems are designed to generate electricity closer to where it is used, often at the scale of individual buildings, neighborhoods, or small communities. **Microgrids**, which are a key component of decentralized energy systems, provide localized control over energy resources and can operate independently or in conjunction with the larger national grid.

In a traditional **centralized energy system**, electricity is produced at large power plants—typically fueled by coal, natural gas, nuclear energy, or hydropower—and transmitted over long distances via high-voltage transmission lines to end users. This model has several limitations, including energy losses during transmission, vulnerability to grid disruptions, and reliance on large, often carbon-intensive energy sources. In contrast, decentralized systems generate power closer to the point of use, often from renewable sources like **solar panels**, **wind turbines**, or **biogas** systems. This approach reduces transmission losses, increases energy security, and enhances resilience.

Microgrids are localized energy systems that can operate autonomously or connect to the broader grid. A microgrid typically consists of **distributed energy resources** (such as solar panels or wind turbines), **energy storage** (such as batteries), and **load management systems** that balance supply and demand within the system. One of the most important features of microgrids is their ability to operate in **island mode**, meaning they can disconnect from the main grid and continue to supply power during outages or periods of grid instability. This makes microgrids an essential tool for increasing **energy resilience**, particularly in regions prone to extreme weather events or natural disasters that can disrupt centralized power systems.

The rise of decentralized energy systems and microgrids is being driven by several factors, including the **growing adoption of renewable energy**, the need for more reliable power in remote or underserved areas, and advancements in **energy storage technology**. As solar panels, wind turbines, and batteries become more affordable and efficient, it is becoming increasingly viable for individuals, businesses, and communities to produce and manage their own energy.

One of the primary benefits of decentralized systems is their ability to support **energy independence**. Homes equipped with rooftop solar panels and battery storage, for example, can generate much of their electricity independently, reducing reliance on the national grid and fossil fuels. In rural or remote areas where building and maintaining grid infrastructure is expensive or impractical, decentralized energy systems offer a sustainable and cost-effective alternative. For instance, in many parts of **Sub-Saharan Africa** and **South Asia**, where millions of people lack access to reliable electricity, decentralized systems powered by solar energy are helping to close the energy access gap and support economic development.

Microgrids also have a critical role in enhancing **grid resilience**. As climate change increases the frequency and severity of extreme weather events, traditional grids are becoming more vulnerable to disruptions. **Hurricanes**, **wildfires**, and **floods** can knock out power to entire regions, sometimes for days or weeks. Microgrids, by contrast, allow communities to maintain power during grid outages, ensuring critical infrastructure—such as hospitals, water treatment plants, and emergency services—can continue operating. This **energy security** is vital for disaster preparedness and response.

In addition to enhancing resilience, decentralized energy systems contribute to **lower greenhouse gas emissions**. By generating electricity from renewable sources like solar, wind, and biomass, microgrids reduce dependence on fossil fuels. Many microgrids are also equipped with **smart technology** that optimizes energy use, helping to reduce overall consumption. For example, a microgrid can manage electricity demand by shifting non-essential energy use to off-peak times or automatically adjusting energy storage to maximize the use of renewable energy.

One of the challenges in adopting decentralized energy systems and microgrids is the need for **advanced infrastructure** and coordination between local systems and the central grid. Integrating distributed resources into the broader grid requires the development of **smart grids**, which use digital technology to monitor and manage energy flows in real-time. This level of connectivity is essential for ensuring that decentralized systems can function efficiently while supporting the stability of the national grid.

Despite these challenges, there is growing interest in decentralized energy systems and microgrids from both **governments** and **private sector** investors. **Regulatory changes** that encourage the development of DERs, such as feed-in tariffs and net metering, are helping to accelerate the adoption of these technologies. In addition, corporate and industrial users are increasingly turning to microgrids to ensure reliable power for their operations and reduce their carbon footprint.

CHAPTER 15: WASTE MANAGEMENT

Types of Waste: Municipal, Industrial, Hazardous

Waste management involves handling various types of waste that are produced by different sectors of society. Understanding the distinctions between **municipal**, **industrial**, and **hazardous waste** is critical for creating effective strategies to manage and minimize environmental harm. Each type of waste comes from different sources and has unique characteristics, requiring specialized treatment, disposal, and recycling methods.

Municipal Waste, often called **municipal solid waste (MSW)**, includes the everyday items discarded by households, offices, schools, and small businesses. This type of waste is what most people think of as **garbage** or **trash**, and it consists of materials such as paper, plastics, glass, food scraps, and yard waste. As urban populations grow, the volume of municipal waste continues to increase, creating significant challenges for waste management systems.

Municipal waste is typically collected by local authorities and either **recycled**, sent to **landfills**, or **incinerated**. Recycling is an essential process for reducing the amount of waste that ends up in landfills, conserving resources, and decreasing pollution. However, many materials, particularly plastics and food waste, still end up in landfills, where they can generate **methane**, a potent greenhouse gas, as they decompose. Modern waste management systems are evolving to address these issues by encouraging **composting**, **waste-to-energy** technologies, and **zero-waste** initiatives.

Industrial Waste comes from the production of goods and services and is generated by manufacturing plants, factories, mining operations, and construction activities. Industrial waste is typically much larger in volume compared to municipal waste and includes a wide range of materials, from **scrap metals** and **chemicals** to **sludge**, **stone**, and **concrete debris**.

Unlike municipal waste, which is relatively homogenous, industrial waste varies greatly depending on the industry. For example, the construction industry generates **debris** like wood, bricks, and concrete, while the chemical industry produces large amounts of **toxic byproducts** that need careful handling. Much of the industrial waste is **non-hazardous**, and materials like metals, plastics, and glass can be recycled. However, industrial waste also includes **e-waste** (electronic waste), which contains harmful substances like **lead** and **mercury** that require specialized disposal.

Industrial waste is often managed on-site or sent to **specialized facilities** for recycling or treatment before disposal. For industries, **sustainable waste management** has become a priority, with many companies adopting strategies like **circular economy** practices, where waste is minimized, and materials are continually reused in production processes.

Hazardous Waste is waste that poses a serious risk to human health or the environment due to its toxicity, flammability, corrosiveness, or reactivity. It includes **chemical waste**, **medical waste**, **radioactive materials**, and **pesticides**. Hazardous waste is generated by both households and industries. For example, households might dispose of batteries, paints, or cleaning agents, while industrial sources produce toxic chemicals or waste from mining and energy production.

One of the key challenges in hazardous waste management is ensuring safe disposal and preventing contamination of water, air, or soil. **Incineration** is often used to dispose of hazardous waste, as it reduces the volume and destroys harmful substances. However, this method must be carefully managed to prevent the release of toxic emissions. **Secure landfills** designed for hazardous waste are another option, but these landfills must be highly regulated to prevent leakage into the surrounding environment.

In addition to incineration and landfilling, **chemical neutralization**, **bioremediation**, and **stabilization** techniques are used to treat hazardous waste and make it less dangerous. The safe transportation and disposal of hazardous materials are strictly regulated by government agencies to minimize the risks to public health and ecosystems.

Waste Disposal Methods: Landfills, Incineration, Recycling

Waste disposal methods have evolved to manage the growing amount of waste generated by households, industries, and businesses. The primary methods used today are **landfills**, **incineration**, and **recycling**. Each approach has its advantages and drawbacks in terms of environmental impact, efficiency, and sustainability.

Landfills are the most common method of waste disposal globally. In this process, waste is buried in the ground, with layers of soil or other material covering it to minimize exposure to the environment. Modern landfills are designed with **liners** and **leachate collection systems** to prevent toxic liquids from seeping into groundwater, and **methane collection systems** are used to capture methane gas produced by decomposing organic waste.

While landfills are a straightforward solution, they come with significant environmental concerns. **Methane**, a potent greenhouse gas, is released as organic waste decomposes in an oxygen-poor environment. Even with methane collection systems, not all of the gas is captured, contributing to **global warming**. Additionally, landfills require large amounts of space, leading to **habitat destruction** and **land use issues**. Over time, improperly managed landfills can leak toxins into the soil and water, causing long-term environmental harm.

Incineration, also known as **waste-to-energy**, involves burning waste at high temperatures to reduce its volume and generate energy. Incineration is particularly useful for disposing of waste that cannot be easily recycled or composted, such as contaminated or hazardous materials. The heat generated from burning waste can be used to produce electricity or heat, making incineration a dual-purpose method of waste management.

However, incineration has its own environmental challenges. Burning waste produces **air pollutants**, including **dioxins, furans**, and **particulate matter**, which can harm human health and the environment if not properly controlled. Modern incinerators use **scrubbers** and **filters** to capture these pollutants, but concerns about toxic emissions remain. Additionally, incineration can discourage waste reduction and recycling efforts, as some waste-to-energy plants rely on a constant supply of waste to operate efficiently.

Recycling is one of the most environmentally friendly waste disposal methods, as it involves converting waste materials into new products. Materials like **paper, glass, metals**, and **plastics** can be collected, processed, and reused, reducing the need for new raw materials. Recycling saves energy, reduces pollution, and conserves natural resources. For example, recycling aluminum saves about **95% of the energy** required to produce aluminum from raw materials.

Despite its benefits, recycling has limitations. Not all materials can be easily recycled, and contamination in recycling streams can reduce the efficiency of the process. **Plastic recycling** is particularly challenging due to the variety of plastic types and their differing recycling requirements. Additionally, many countries face challenges in managing **e-waste** (electronic waste), as improper recycling can release toxic substances like lead and mercury into the environment.

Each waste disposal method—landfills, incineration, and recycling—has a role in managing waste, but none is a perfect solution on its own. A combination of waste management strategies, supported by **waste reduction** efforts and innovations in recycling technology, is necessary to minimize environmental impacts and create a more sustainable system.

The Circular Economy and Waste Reduction

The concept of a **circular economy** offers a transformative approach to waste reduction by rethinking how we produce, consume, and dispose of goods. Unlike the traditional **linear economy**, which follows a "take-make-dispose" model, the circular economy seeks to close the loop by keeping materials in use for as long as possible. This approach not only reduces waste but also conserves resources, lowers emissions, and promotes sustainability.

At the heart of the circular economy is the idea that products should be designed for **reuse**, **repair**, **remanufacturing**, and **recycling**. Instead of being discarded after their useful life, materials are recovered and reintegrated into the production process. For example, in a circular economy, a product like a washing machine would be designed so that its components can be easily repaired or replaced, extending its lifespan. When it eventually reaches the end of its life, the machine's parts would be recovered and used to manufacture new appliances, rather than ending up in a landfill.

One of the key strategies for promoting a circular economy is **waste reduction at the source**. By designing products that use fewer materials or that can be easily disassembled, manufacturers can minimize the amount of waste generated. This approach is known as **eco-design** and is being adopted by many industries. For instance, companies are reducing packaging waste by using **biodegradable** or **reusable materials**, and manufacturers are creating electronics that are easier to repair and upgrade rather than replace.

Another important aspect of the circular economy is **remanufacturing**, where products are restored to like-new condition using a combination of reused, repaired, and new parts. This process is particularly common in industries like **automobile manufacturing**, where remanufactured parts can be a cost-effective and environmentally friendly alternative to new parts. **Product life extension**, achieved through repair services or upgrades, also has a key part in reducing waste by ensuring that products stay in use longer.

Recycling is a central component of the circular economy, but the goal is to move beyond recycling as a last resort and focus more on **preventing waste** in the first place. When recycling is necessary, it is optimized to ensure that materials are recovered at their highest value. In a fully realized circular economy, waste from one industry becomes the raw material for another. This is known as **industrial symbiosis**, where companies work together to exchange byproducts, turning waste into valuable resources.

For consumers, the circular economy promotes more conscious consumption patterns. This means choosing **durable products**, supporting companies that prioritize **sustainable production**, and embracing alternatives like **product-as-a-service models**. In these models, consumers lease or rent products rather than owning them outright, which encourages manufacturers to design long-lasting, repairable goods.

Governments and businesses are increasingly recognizing the benefits of a circular economy, not just for waste reduction but also for **economic growth** and **job creation**. By keeping materials in circulation, the circular economy reduces the need for virgin resource extraction, lowers greenhouse gas emissions, and cuts energy consumption. It also stimulates innovation in product design, material science, and recycling technologies.

Transitioning to a circular economy requires collaboration between governments, industries, and consumers. Through **policy initiatives**, such as **extended producer responsibility (EPR)**, governments can encourage companies to take responsibility for the entire lifecycle of their products, from design to disposal. At the same time, consumers can drive demand for more sustainable products and services by making environmentally conscious purchasing decisions.

Composting and Organic Waste Solutions

Composting is a natural process of recycling organic waste into nutrient-rich soil, providing an environmentally friendly solution to manage organic waste. Organic waste includes food scraps, yard trimmings, and agricultural residues that would otherwise end up in landfills, contributing to methane emissions. By diverting this waste through composting, we can reduce landfill usage, decrease greenhouse gas emissions, and enrich soil health, creating a sustainable loop for organic material.

Composting works by allowing organic materials to decompose under controlled conditions. Microorganisms such as bacteria and fungi break down organic matter into **humus**, a dark, nutrient-rich material. The process requires three essential elements: **carbon**, provided by dry materials like leaves or straw; **nitrogen**, found in green materials like food scraps or grass clippings; and **oxygen**, which is necessary to fuel the decomposition process. Properly managed compost piles are regularly turned to ensure air circulation and moisture balance, which helps the organic matter decompose efficiently and without producing unpleasant odors.

There are several types of composting systems that cater to different needs and scales. **Backyard composting** is the simplest and most common method, where households can compost kitchen scraps and yard waste in a small bin or heap. Larger-scale solutions include **community composting**, which serves multiple households or neighborhoods, and **industrial composting**, which processes organic waste on a much larger scale, often through municipal waste programs. **Vermicomposting**, which uses worms to break down organic material, is another option for households with limited space.

One of the major benefits of composting is that it reduces the amount of waste sent to landfills. In many countries, organic waste makes up a significant portion of municipal solid waste. When this waste decomposes in landfills, it generates

methane, a potent greenhouse gas that contributes to climate change. Composting helps mitigate this by converting organic matter into a valuable product instead of allowing it to break down anaerobically in a landfill.

Composting also enhances **soil health** by improving soil structure, increasing water retention, and providing essential nutrients for plant growth. The resulting compost enriches soil with nitrogen, phosphorus, and potassium, all vital for healthy crops and plants. This can reduce the need for synthetic fertilizers, which are often derived from fossil fuels and contribute to environmental pollution when overused.

In urban areas, **food waste** is a significant source of organic material that often ends up in landfills. Many cities are implementing **food waste collection programs** where households and businesses can separate organic waste for composting. By scaling up composting initiatives and promoting **waste-to-soil programs**, municipalities can reduce the burden on landfills, lower waste disposal costs, and provide a steady supply of compost for local parks, gardens, and agriculture.

Composting represents an effective, low-cost solution for managing organic waste. It not only diverts material from landfills but also contributes to healthier soils, reduces the need for chemical fertilizers, and lowers methane emissions. As more individuals, communities, and governments embrace composting, it becomes a powerful tool in building a more sustainable and resilient waste management system.

Plastic Waste and Ocean Pollution

Plastic waste has become one of the most pressing environmental issues of the 21st century, particularly in relation to **ocean pollution**. Every year, millions of tons of plastic enter the world's oceans, where it harms marine life, disrupts ecosystems, and contributes to environmental degradation. Plastic, especially single-use items like bottles, bags, and straws, does not decompose easily. Instead, it breaks down into smaller particles known as **microplastics**, which persist in the environment for centuries.

The majority of plastic waste in the ocean originates from land-based sources, such as **poor waste management practices**, **littering**, and **runoff** from urban areas. Inadequate recycling infrastructure, particularly in developing countries, means that much of the plastic waste generated ends up in rivers and coastal areas, where it is carried out to sea. **Fishing gear**, such as nets and lines, also contributes significantly to plastic pollution in the ocean, with **ghost nets** continuing to trap marine life long after being discarded.

Once in the ocean, plastic waste has devastating effects on marine ecosystems. **Marine animals**, including fish, sea turtles, and seabirds, often mistake plastic debris for food, leading to ingestion, choking, or starvation. For example, sea turtles often mistake plastic bags for jellyfish, one of their primary food sources. Consuming plastic can block their digestive tracts, leading to malnutrition or death. Similarly, seabirds and fish that consume small plastic fragments accumulate toxins in their bodies, which can move up the food chain, potentially affecting humans who consume contaminated seafood.

Microplastics, which are smaller than 5 millimeters, pose a unique challenge. These tiny particles are now pervasive in marine environments, having been found from coastal waters to the deepest parts of the ocean. Microplastics can originate from the breakdown of larger plastic items, or from products like **microbeads** found in personal care products and **synthetic fibers** from clothing. Once in the water, microplastics can be ingested by marine organisms, from plankton to whales, causing internal damage and exposing them to harmful chemicals.

Plastic waste also harms **marine habitats** such as coral reefs, mangroves, and seagrass beds. Plastic debris can smother coral reefs, blocking sunlight and hindering the growth of corals. Additionally, as plastics degrade, they release toxic chemicals, including **bisphenol A (BPA)** and **phthalates**, which can disrupt the hormonal systems of marine animals, affecting their growth, reproduction, and survival.

One of the most visible manifestations of ocean plastic pollution is the formation of **garbage patches** in the world's oceans, the largest being the **Great Pacific Garbage Patch**. These patches are not solid islands of trash, but rather vast areas of floating plastic debris dispersed by ocean currents. The presence of these patches highlights the global nature of plastic pollution, as plastic waste from different countries converges in these gyres.

Efforts to address plastic pollution focus on both **prevention** and **clean-up**. Reducing the production and use of single-use plastics is a critical step. Many countries and cities have implemented **plastic bag bans, straw bans,** or taxes to encourage consumers to shift to reusable alternatives. **Recycling** systems also need to be improved to prevent plastic waste from entering the environment in the first place. However, less than 10% of the world's plastic is currently recycled, underscoring the need for a more effective global approach to plastic waste management.

In terms of ocean clean-up, organizations like **The Ocean Cleanup** are developing technologies to remove plastic from the ocean, focusing on intercepting plastic in rivers before it reaches the sea. However, clean-up efforts alone are not enough to address the scale of the problem. Combating plastic pollution requires a coordinated effort that includes reducing plastic production, improving waste management, and raising public awareness about the impacts of plastic on the environment.

CHAPTER 16: ENVIRONMENTAL POLICY AND REGULATION

Environmental Laws: National and International

Environmental laws at both the national and international levels serve to protect natural resources, control pollution, and ensure the sustainable use of ecosystems. These laws set standards for air, water, and soil quality, regulate emissions, and preserve biodiversity. The frameworks vary across regions but share the goal of addressing environmental challenges through legal mechanisms and policies.

At the **national level**, many countries have enacted comprehensive environmental legislation to manage pollution and protect ecosystems. In the United States, the **Clean Air Act (CAA)** is a landmark piece of legislation that regulates air emissions from both stationary and mobile sources. Enacted in 1970 and amended in 1990, the CAA authorizes the **Environmental Protection Agency (EPA)** to establish **National Ambient Air Quality Standards (NAAQS)** to protect public health and the environment from harmful pollutants like sulfur dioxide, nitrogen oxides, and particulate matter. The law also includes provisions to limit emissions of hazardous air pollutants and to phase out chemicals that contribute to **ozone depletion**.

The **Clean Water Act (CWA)**, another foundational U.S. environmental law, regulates discharges of pollutants into the nation's waterways. Passed in 1972, the CWA aims to restore and maintain the chemical, physical, and biological integrity of U.S. waters. It sets water quality standards for contaminants and requires industries to obtain permits before discharging pollutants into rivers, lakes, and streams.

In the European Union, the **EU Environmental Action Programmes** outline long-term environmental strategies. The **European Green Deal** is a key policy framework aiming to make Europe climate-neutral by 2050. This includes laws targeting **carbon emissions**, **energy efficiency**, and **biodiversity conservation**. The EU also enforces the **Water Framework Directive**, which sets goals for improving water quality and managing water resources sustainably across member states.

Environmental laws also address **hazardous waste management**. In many countries, regulations like the U.S. **Resource Conservation and Recovery Act (RCRA)** govern the handling, storage, and disposal of hazardous materials. RCRA also emphasizes the need for reducing waste generation and encourages recycling and reuse.

On the **international stage**, environmental laws and agreements are essential for addressing global challenges like climate change, ozone depletion, and biodiversity loss. The **Paris Agreement**, adopted in 2015 under the **United Nations**

Framework Convention on Climate Change (UNFCCC), is one of the most significant international treaties. The agreement aims to limit global warming to well below 2°C above pre-industrial levels, with efforts to keep it below 1.5°C. Under this framework, countries set their own **nationally determined contributions (NDCs)** to reduce greenhouse gas emissions and are expected to report on their progress regularly.

The **Montreal Protocol** on substances that deplete the ozone layer is another successful international environmental treaty. Adopted in 1987, the protocol targets the phase-out of **chlorofluorocarbons (CFCs)** and other ozone-depleting chemicals. The Montreal Protocol is widely regarded as one of the most effective environmental agreements, having led to the near elimination of CFCs and the gradual recovery of the ozone layer.

The **Convention on Biological Diversity (CBD)**, signed in 1992 at the **Earth Summit** in Rio de Janeiro, addresses the conservation of biodiversity, sustainable use of its components, and fair sharing of genetic resources. It sets goals for countries to protect ecosystems, species, and genetic diversity through national strategies and action plans.

Other key international agreements include the **Basel Convention**, which regulates the transboundary movement of hazardous waste and aims to prevent illegal dumping in developing countries, and the **Kyoto Protocol**, an earlier agreement focused on reducing greenhouse gas emissions, which has now been largely succeeded by the Paris Agreement.

National and international environmental laws work together to address both local and global environmental issues. National laws provide the framework for implementing policies that protect air, water, and land resources, while international agreements enable countries to collaborate on transboundary environmental challenges, from climate change to pollution control. Together, they form a comprehensive legal approach to safeguarding the environment for future generations.

Role of Governments in Environmental Protection

Governments are important in environmental protection by creating and enforcing laws, regulations, and policies designed to preserve natural resources, reduce pollution, and safeguard ecosystems. Their actions are critical for addressing complex environmental challenges that individual actions or market mechanisms alone cannot solve, such as air and water pollution, climate change, and habitat destruction. Through legislation, regulation, and international cooperation, governments provide the framework for environmental sustainability.

One of the primary ways governments contribute to environmental protection is through **environmental legislation**. Laws such as the **Clean Air Act**, **Clean Water Act**, and **Endangered Species Act** in the United States, or the **Environmental Protection Act** in other countries, set specific standards for pollution control, natural resource use, and wildlife conservation. These laws often establish limits on emissions of pollutants, regulate waste management practices, and protect critical habitats from development or exploitation. Enforcement of these laws is carried out by agencies such as the **Environmental Protection Agency (EPA)** in the U.S., or similar bodies in other countries, which monitor compliance and take action against violators.

In addition to national legislation, governments have a key role in setting **international environmental standards** through participation in global agreements. Treaties like the **Paris Agreement** on climate change or the **Montreal Protocol** on ozone depletion are negotiated by governments to address issues that cross national boundaries. These international agreements require cooperation among countries to reduce emissions, phase out harmful chemicals, or protect endangered species. By joining such treaties, governments commit to meeting specific environmental goals and implementing policies that contribute to global sustainability.

Governments also promote environmental protection through **economic incentives and disincentives**. For example, they may introduce **carbon taxes** or **cap-and-trade systems** to reduce greenhouse gas emissions, effectively making it more costly for businesses to pollute while encouraging investment in cleaner technologies. **Subsidies for renewable energy** sources like solar, wind, and geothermal power provide financial incentives for businesses and consumers to shift away from fossil fuels. Additionally, tax credits or rebates for energy-efficient appliances and electric vehicles help reduce overall energy consumption and promote sustainable practices among the public.

Land-use planning and conservation programs are another important area where governments intervene to protect the environment. Through zoning laws, protected areas, and national parks, governments can restrict activities that harm the environment, such as deforestation, overfishing, or industrial development in sensitive areas. The creation of **wildlife reserves** or **marine protected areas** ensures that biodiversity is preserved, allowing ecosystems to thrive while providing important services like carbon sequestration, water purification, and soil preservation.

Regulatory agencies are crucial in implementing and enforcing environmental laws. Agencies like the **EPA, European Environment Agency (EEA)**, or **National Environmental Management Authority (NEMA)** in various countries are responsible for developing regulations based on environmental laws, conducting environmental assessments, issuing permits, and enforcing compliance. These agencies also gather data on environmental conditions, monitor pollution levels, and respond to environmental emergencies such as oil spills or industrial accidents.

Governments are also involved in **public education and awareness campaigns**. Informing citizens about environmental issues, such as climate change, biodiversity loss, or waste reduction, is an important function of government agencies. By raising awareness, governments help foster more sustainable behaviors, such as reducing plastic use, conserving energy, or protecting wildlife habitats.

While governments are pivotal in setting the regulatory framework, they must also balance competing interests. Businesses, industry groups, and political stakeholders often lobby against strict environmental regulations, citing potential economic costs or job losses. Effective environmental protection requires governments to navigate these pressures while ensuring that long-term sustainability goals are met.

Ultimately, governments act as the primary stewards of national and international environmental policy, using legislation, regulation, economic tools, and public education to guide society toward more sustainable practices. Their role is essential in ensuring that natural resources are protected and that environmental damage is minimized for future generations.

Non-Governmental Organizations and Activism

Non-governmental organizations (NGOs) and activism are important in environmental protection by raising awareness, advocating for policy changes, and holding governments and corporations accountable. These organizations operate independently of governments and often focus on specific environmental issues such as climate change, deforestation, wildlife conservation, and pollution.

NGOs like **Greenpeace**, the **World Wildlife Fund (WWF)**, and **Friends of the Earth** work globally and locally to protect the environment through a combination of advocacy, research, and public engagement. They monitor environmental conditions, conduct scientific studies, and publish reports that highlight pressing issues like habitat destruction, species extinction, and the impacts of industrial pollution. These reports help inform the public and policymakers, pushing for stricter environmental regulations or changes in corporate behavior.

Activism, often led by environmental NGOs, involves direct action such as protests, petitions, and campaigns to draw attention to environmental issues. Movements like **Fridays for Future**, inspired by climate activist **Greta Thunberg**, have mobilized millions of people worldwide to demand stronger climate action from governments and industries. These movements amplify public pressure, making it harder for political leaders to ignore environmental concerns.

NGOs also engage in **legal actions**, suing governments or corporations for violating environmental laws. For example, organizations like **Earthjustice** provide legal representation to communities affected by environmental degradation, using the court system to enforce regulations or challenge decisions that harm the environment.

Through their advocacy, research, and public engagement, NGOs and activists serve as watchdogs, ensuring that environmental issues remain high on the global agenda. They help shape public opinion, influence policy, and drive meaningful environmental reforms.

Public Participation in Environmental Decision-Making

Public participation in environmental decision-making is a key principle of democratic governance, ensuring that citizens have a voice in shaping policies that affect their environment. It allows communities to be actively involved in decisions about land use, resource management, and environmental protection, fostering transparency and accountability in government actions.

Public involvement typically begins with **consultation processes**. When governments or companies propose projects that could impact the environment— such as building infrastructure, developing natural resources, or approving industrial facilities—they are often required to consult the public. This process can take the form of public hearings, town hall meetings, or comment periods where individuals and communities can express their views and concerns. For instance, when a new highway or dam is proposed, affected communities can provide feedback on the potential environmental impacts, suggesting alternatives or voicing opposition if the project threatens ecosystems or public health.

Environmental Impact Assessments (EIAs) are another tool for public participation. EIAs are required for major projects, and they include opportunities for the public to review the assessments and provide input. By participating in EIAs, citizens can influence decisions on projects like mining operations, logging activities, or energy development, ensuring that environmental risks are fully considered before approval.

Stakeholder engagement is essential when addressing issues like pollution, habitat loss, or climate adaptation strategies. Governments often convene **multi-stakeholder platforms** that include representatives from affected communities, businesses, environmental organizations, and local authorities. These platforms provide a forum for collaborative decision-making, where different interests are weighed and considered before implementing environmental policies. This approach is particularly important for decisions that involve indigenous communities or rural populations that rely on natural resources for their livelihoods.

Public participation is also critical in shaping **environmental policies and regulations**. In many countries, draft laws or regulations are subject to public comment before being finalized. This allows individuals, environmental groups, and businesses to provide input, ensuring that laws are balanced and consider the concerns of all stakeholders. For example, when new air quality standards are proposed, public input can help refine the regulations to better protect vulnerable populations, such as children or those with respiratory conditions.

Beyond formal processes, the public can participate in environmental decision-making through **grassroots activism** and **community-led initiatives**. Local organizations often form in response to environmental issues such as pollution, deforestation, or water contamination, empowering communities to take direct action. These groups engage in activities like restoring ecosystems, monitoring pollution, or organizing campaigns to protect natural areas. By working together, communities can advocate for stronger protections and push back against environmentally harmful practices.

The use of **digital platforms** has expanded public participation in recent years. Online petitions, social media campaigns, and virtual town halls enable broader engagement, allowing more people to participate in environmental decision-making regardless of geographic location. These tools have made it easier for citizens to voice their opinions and mobilize around environmental causes.

Public participation strengthens environmental decision-making by ensuring that the voices of those most affected are heard and considered. It promotes more equitable and informed outcomes, helping to build consensus around sustainable solutions and fostering a sense of ownership over environmental protection efforts.

APPENDIX

Terms and Definitions

- **Abiotic** – Non-living physical and chemical elements in an ecosystem, like water, air, and minerals.
- **Acid Rain** – Precipitation with a pH lower than normal, caused by sulfur dioxide and nitrogen oxides in the atmosphere.
- **Aerobic Respiration** – The process in which organisms convert glucose and oxygen into energy, releasing carbon dioxide and water.
- **Albedo** – The reflectivity of a surface; higher albedo surfaces (like ice) reflect more sunlight.
- **Aquifer** – An underground layer of rock or sediment that holds water and allows it to flow.
- **Biodiversity** – The variety of life on Earth, encompassing species diversity, genetic diversity, and ecosystem diversity.
- **Biodegradable** – A substance capable of being decomposed by microorganisms, reducing its environmental impact.
- **Bioaccumulation** – The gradual buildup of chemicals or toxins in an organism over time.
- **Biogeochemical Cycles** – The movement of elements and compounds like carbon and nitrogen through living organisms and the environment.
- **Biotic** – Living components of an ecosystem, such as plants, animals, and microorganisms.
- **Cap and Trade** – A market-based system that allows countries or companies to trade emissions allowances to reduce overall greenhouse gas emissions.
- **Carbon Cycle** – The movement of carbon between the atmosphere, biosphere, oceans, and geosphere.
- **Carbon Sequestration** – The process of capturing and storing atmospheric carbon dioxide to reduce climate change.
- **Carcinogen** – A substance capable of causing cancer in living tissue.
- **Carrying Capacity** – The maximum number of individuals that an environment can support without degradation.
- **Climate** – The average weather conditions of an area over a long period of time.
- **Climate Change** – Long-term changes in temperature and weather patterns, often linked to human activities.
- **Composting** – The process of breaking down organic waste into nutrient-rich soil amendment.
- **Conservation** – The sustainable management of natural resources to prevent exploitation, destruction, or neglect.
- **Deforestation** – The removal of forests, often leading to habitat loss, biodiversity decline, and climate impacts.
- **Desertification** – The process by which fertile land becomes desert, often as a result of drought, deforestation, or inappropriate agriculture.

- **Ecological Footprint** – A measure of the human demand on Earth's ecosystems, representing the amount of natural resources consumed.
- **Ecology** – The study of interactions between organisms and their environment.
- **Ecosystem** – A community of living organisms interacting with their physical environment.
- **Ecotourism** – Responsible travel to natural areas that conserves the environment and supports local communities.
- **Endangered Species** – Species at risk of extinction due to loss of habitat, poaching, or environmental changes.
- **Endemic Species** – Species that are native to a specific geographic area and found nowhere else in the world.
- **Energy Efficiency** – Using less energy to perform the same task, reducing energy waste.
- **Environmental Degradation** – The deterioration of the environment through depletion of resources, destruction of ecosystems, and pollution.
- **Environmental Ethics** – The moral relationship between humans and the natural world, often focusing on sustainability and conservation.
- **Eutrophication** – Excessive nutrient enrichment in water bodies, leading to oxygen depletion and harm to aquatic life.
- **Ex-Situ Conservation** – Conservation of species outside their natural habitat, such as in zoos or botanical gardens.
- **Fossil Fuels** – Energy sources like coal, oil, and natural gas that are formed from ancient organic matter.
- **Geoengineering** – Large-scale intervention strategies designed to mitigate the effects of climate change, such as altering the atmosphere or oceans.
- **Greenhouse Effect** – The warming of Earth's atmosphere due to the trapping of heat by greenhouse gases like CO_2 and methane.
- **Habitat Fragmentation** – The division of larger, contiguous habitats into smaller, isolated sections, often due to human activities.
- **Hydrological Cycle** – The movement of water through the environment via processes like evaporation, condensation, and precipitation.
- **In-Situ Conservation** – The conservation of species in their natural habitats, maintaining ecosystem balance.
- **Indicator Species** – Species whose presence, absence, or abundance reflects the overall health of an ecosystem.
- **Invasive Species** – Non-native species that spread rapidly in new environments and disrupt ecosystems.
- **Keystone Species** – A species that has a critical role in maintaining the structure of an ecosystem.
- **Landfill** – A site for the disposal of waste material by burying it under layers of earth.
- **Microplastics** – Tiny plastic particles that result from the breakdown of larger plastics, which can accumulate in marine and freshwater ecosystems.
- **Mitigation** – Measures taken to reduce or prevent the effects of environmental damage.
- **Natural Capital** – The world's stocks of natural resources, including geology, soil, air, water, and living organisms.

- **Natural Selection** – The process by which organisms better adapted to their environment tend to survive and reproduce.
- **Non-Renewable Resources** – Resources like fossil fuels that are finite and not replenished on a human timescale.
- **Overexploitation** – The excessive use of resources or species to the point of depletion or extinction.
- **Ozone Layer** – A layer in the Earth's stratosphere containing high concentrations of ozone, which protects the planet from harmful UV radiation.
- **Photosynthesis** – The process by which green plants convert light energy into chemical energy, producing oxygen and glucose.
- **Pollutant** – A substance introduced into the environment that has harmful or poisonous effects.
- **Pollution** – The contamination of the natural environment by harmful substances or waste.
- **Renewable Resources** – Natural resources that can be replenished over time, such as solar energy, wind, and water.
- **Restoration Ecology** – The scientific study and practice of restoring ecosystems to their natural state.
- **Smog** – A type of air pollution caused by the interaction of sunlight with pollutants like vehicle emissions and industrial gases.
- **Species Richness** – The number of different species represented in an ecological community or ecosystem.
- **Stewardship** – The responsible management and care of natural resources and ecosystems.
- **Sustainable Agriculture** – Farming practices that maintain the long-term health of the soil, ecosystems, and environment.
- **Sustainable Development** – Development that meets the needs of the present without compromising the ability of future generations to meet theirs.
- **Symbiosis** – A relationship between two species where at least one benefits; includes mutualism, commensalism, and parasitism.
- **Trophic Levels** – The hierarchical levels in an ecosystem, each representing a step in the flow of energy and nutrients.
- **Tundra** – A cold, treeless biome with low-growing vegetation, found in the Arctic and Antarctic regions.
- **Urbanization** – The process of human populations increasing in cities, often leading to environmental strain.
- **Water Table** – The level below which the ground is saturated with water.
- **Watershed** – An area of land where all the water drains into a common waterway, such as a river or lake.
- **Wetlands** – Ecosystems where water covers the soil for significant periods of time, providing habitats for diverse species.
- **Zonation** – The spatial arrangement of ecosystems or species in response to environmental gradients like altitude, moisture, or light.

AFTERWORD

Congratulations! By reaching this point, you've gained a broad understanding of our planet's environmental systems and the challenges we face in preserving them.

Looking Back

When we started this book, we began with the fundamentals of environmental science. Step by step, we explored ecosystems, the cycling of elements through Earth's systems, the delicate balance of our atmosphere, and the incredible diversity of life on our planet. We've examined the impacts of human activities on these systems and discovered the innovative solutions being developed to address environmental challenges.

What Comes Next?

Knowledge is just the beginning. As you close this book, you might be wondering, "What can I do with everything I've learned?" The answer is: quite a lot! Whether you're planning to pursue a career in environmental science, looking to make more sustainable choices in your daily life, or hoping to educate others, you now have a solid foundation to build upon.

Environmental science is a constantly evolving field. New discoveries are made every day, and our understanding of Earth's systems continues to grow. Stay curious, keep learning, and don't be afraid to question and explore.

A Personal Note

My hope is that this book has not only provided facts but also inspired you. Environmental challenges can sometimes seem overwhelming, but understanding them is the first step toward solving them. Every positive change, no matter how small, contributes to a larger impact.

As you move forward, keep in mind that you're now part of a global community of environmentally aware individuals. Your knowledge and actions can influence others and contribute to positive change. Whether it's making sustainable choices, participating in conservation efforts, or simply sharing what you've learned with others, you have the power to make a difference.

The Journey Continues

This book may be ending, but your journey with environmental science is just beginning. Keep exploring, keep questioning, and keep working toward a more sustainable future. Our planet needs informed, passionate individuals like you.

Made in the USA
Columbia, SC
11 June 2025

ca7482da-2242-4c53-8a76-de17ba7bc3a0R01